Matthew J. Palys, Mahdi Malmali, Michael Reese and Prodromos Daoutidis
Ammonia for Sustainable Energy and Agriculture

Also of Interest

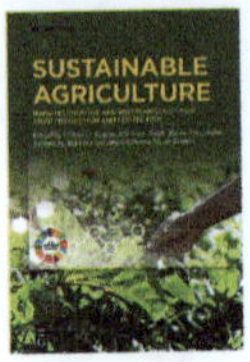

Sustainable Agriculture
Nanotechnology and Biotechnology for Crop Production and Protection
Rajput, Singh, Gharayan, Minkina, Al-Tawaha (Eds.), 2024
ISBN 978-3-11-123467-0, e-ISBN (PDF) 978-3-11-123469-4,
e-ISBN (EPUB) 978-3-11-123501-1

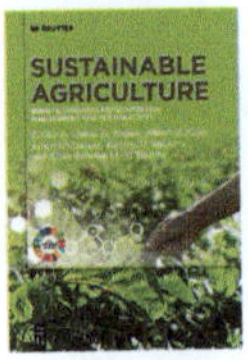

Sustainable Agriculture
Nanotechnology, Biotechnology, Management and Food Security
Rajput, Singh, Gharayan, Minkina, Al-Tawaha (Eds.), 2024
ISBN 978-3-11-136197-0, e-ISBN (PDF) 978-3-11-136975-4,
e-ISBN (EPUB) 978-3-11-136994-5

Plant Protection
From Chemicals to Biologicals
Soni, Suyal, Goel (Eds.), 2022
ISBN 978-3-11-077147-3, e-ISBN (PDF) 978-3-11-077155-8,
e-ISBN (EPUB) 978-3-11-077163-3

Crop Nutrition
Enhancing Healthy Soils, Food Security, Environmental Sustainability and Advancing SDGs
Amanullah, 2024
ISBN 978-3-11-161709-1, e-ISBN (PDF) 978-3-11-161767-1,
e-ISBN (EPUB) 978-3-11-161819-7

Engineered Nanoparticles in Agriculture
From Laboratory to Field
Fotopoulos, Gohari (Eds.), 2023
ISBN 978-1-5015-2315-1, e-ISBN (PDF) 978-1-5015-2322-9,
e-ISBN (EPUB) 978-1-5015-1552-1

Matthew J. Palys, Mahdi Malmali, Michael Reese
and Prodromos Daoutidis

Ammonia for Sustainable Energy and Agriculture

Science, Technology and Economics

DE GRUYTER

Authors
Matthew J. Palys
University of Minnesota
Department of Chemical Engineering
and Materials Science
421 Washington Avenue SE
Minneapolis MN 55455
United States of America

Mahdi Malmali
Texas Tech University
Department of Chemical Engineering
P.O. Box 43121
Lubbock TX 79409
United States of America

Michael Reese
University of Minnesota
West Central Research and Outreach Center
46352 State Highway 329
Morris MN 56267
United States of America

Prodromos Daoutidis
University of Minnesota
Department of Chemical Engineering
and Materials Science
421 Washington Avenue SE
Minneapolis MN 55455
United States of America

ISBN 978-3-11-076247-1
e-ISBN (PDF) 978-3-11-076252-5
e-ISBN (EPUB) 978-3-11-076258-7

Library of Congress Control Number: 2025943006

Bibliographic information published by the Deutsche Nationalbibliothek
The Deutsche Nationalbibliothek lists this publication in the Deutsche Nationalbibliografie; detailed bibliographic data are available on the Internet at http://dnb.dnb.de.

Cover image: i-Stockr / iStock / Getty Images Plus
Typesetting: VTeX UAB, Lithuania
Printing and binding: CPI books GmbH, Leck

www.degruyterbrill.com
Questions about General Product Safety Regulation:
productsafety@degruyterbrill.com

Contents

1 Introduction: Why renewable ammonia?

Since the dawn of the 20th century, industrial agriculture has revolutionized society, allowing greater access to less expensive food. Synthetic ammonia is arguably the backbone of modern agriculture. In addition to water, nitrogen fertilizer is generally considered the second most limiting nutrient for traditional crops such as corn and small grains [96]. The use of ammonia either applied directly or as a precursor to other fertilizers is estimated to provide the nitrogen nutrients that ensure the food supply for approximately 50 % of the global population [344, 91]. Figure 1.1 shows an unmistakable increase in the rate of population growth around 1950, coinciding with a more widespread production of fertilizers based on synthetic ammonia.

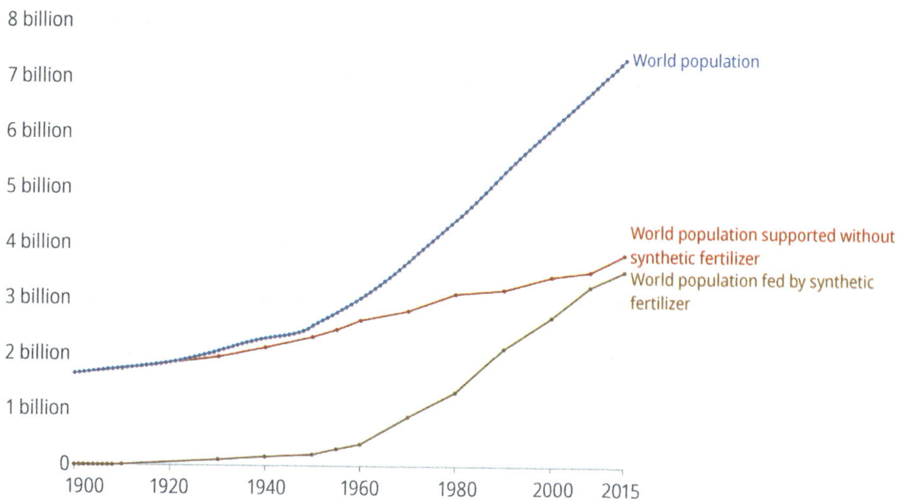

Figure 1.1: Estimates of the global population reliant on synthetic nitrogen fertilizers for food production for which precursor ammonia is produced via the Haber–Bosch process. Best estimates project that just over half of the global population could be sustained without synthetic nitrogen fertilizers. The data in this figure is from [91, 345] and processed by Our World in Data. Reprinted under Creative Common CC BY License.

Nitrogen fertilizers derived from ammonia are used on every continent to increase crop yields. This can allow less land to be used for a given amount of food production or can allow higher levels of food production in countries with high population densities or rapidly growing populations. Figure 1.2 shows the amount of nitrogen fertilizer used per hectare of cropland in each country. Some countries with the highest fertilization rates,

https://doi.org/10.1515/9783110762525-001

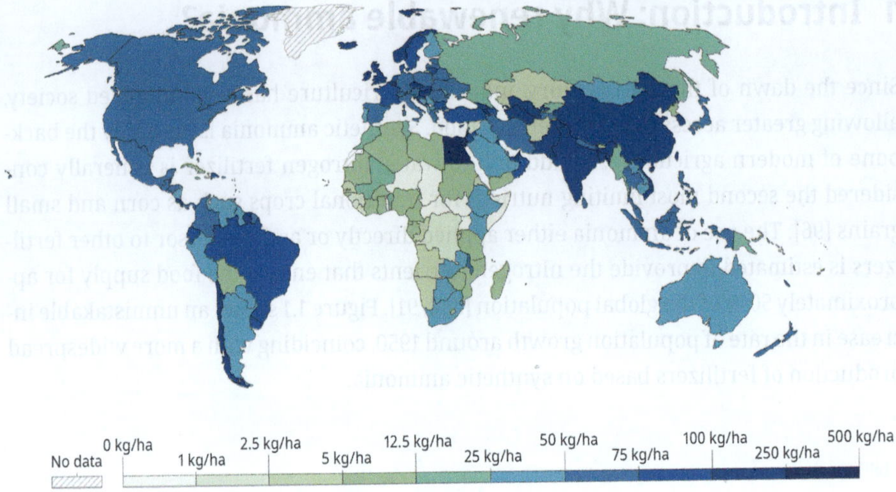

Figure 1.2: Application of nitrogen fertilizer in each country, measured in kilograms of total nutrient per hectare of cropland. The data in this figure is from [100] and processed by Our World in Data. Reprinted under Creative Common CC BY license.

such as India, China, Pakistan, Bangladesh, and Brazil, are among the ten most populous in the world. Others in western Europe, such as the Netherlands and Belgium, have high population densities. Another reason for the high intensities of nitrogen fertilizer application is the low availability of arable land, as is the case in Egypt, the United Arab Emirates, Norway, and Iceland. At the same time, an increase in nitrogen fertilizer availability would provide greater access to food in many African nations. Clearly, synthetic ammonia is essential for modern civilization. In fact, ammonia plays such a significant role that the Haber–Bosch process used for its production has been considered by some to be one of the top five innovations in modern human history. State-of-the-art ammonia production facilities have been developed and optimized over the past 100 years, and today it is the second most synthesized chemical in the world [109].

Ammonia production, trade, and distribution is a global industry. Approximately 200 million metric tons of ammonia were produced in 2024 [5]. This occurred in approximately 450 to 500 production facilities. The demand for ammonia is projected to grow at an average rate of 3.5 % per year until 2040 [5]. Almost 80 % of this ammonia will be used in the agricultural sector to make nitrogen fertilizers such as urea, urea ammonia nitrate (UAN), ammonium nitrate, and ammonium sulfate [401, 1]. Ammonia can also be combined with phosphorus, another important nutrient for crop production, to produce monoammonia and diammonium phosphate (MAP and DAP) fertilizers. Specifically, in 2024, 46 % of global ammonia production was used to make urea, 9 % was used to make ammonium nitrate and its derivatives, 6 % was used to make mono and diammonium phosphates, 15 % was used to make all other nitrogen fertilizers, and 2 % was applied directly to crops as anhydrous ammonia. The other 22 % was used in industrial applications, including use as a refrigerant, as well as in the production of industrial grade urea, commonly used as exhaust fuel in diesel engines, and technical grade ammonium nitrate, commonly used to produce explosives. The production of urea and other commodities is almost exclusively integrated within the same facilities that produce ammonia.

In 2023, China produced the majority of the ammonia produced in the world at approximately 43 million mt. The United States, Russia, and India each produced approximately 14 million mt [21]. Indonesia produced 6 million mt and 17 more countries produced between 1 and 5 million mt. These 22 countries produced 93 % of the world's ammonia. Some of this production is based in countries with the highest population, intended for domestic use, but ammonia and its derivatives are also commodities traded globally. More than 120 ports around the world are equipped with ammonia terminals [3]. This port network as of 2017 is shown in Figure 1.3. The five largest ammonia exporting countries are Trinidad and Tobago, Saudi Arabia, Indonesia, the United States, and Canada. These countries, especially Trinidad and Tobago and Saudi Arabia, have ample natural gas reserves and have located most of this production close to ports with an eye on exports. Meanwhile, the five largest ammonia importing countries by volume in order are India, the United States, Morocco, South Korea, and Turkey. India, South Korea, and Turkey import ammonia primarily to maintain their high and growing populations,

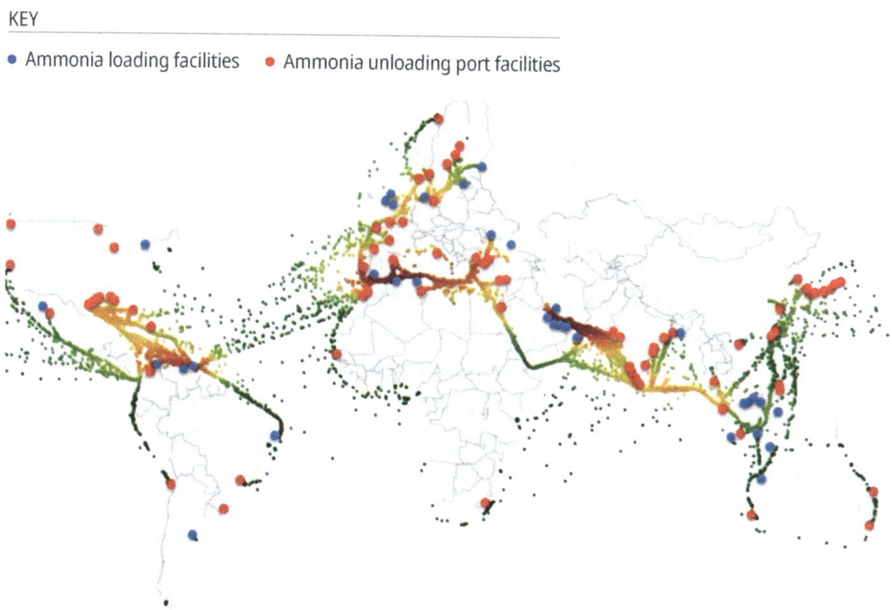

Figure 1.3: Ammonia maritime shipping infrastructure with loading and unloading facilities. Includes heat map with darker red colors showing higher amount of ammonia transportation. Reprinted from [3] under Creative Common CC BY license.

whereas Morocco's imports are predominantly used to produce a majority of the world's ammonium phosphates due to their phosphate reserves. The United States being a top exporter and importer are emblematic of their presence in multinational ammonia and fertilizer production corporations. For example, an American company, Koch Industries, owns equity in multiple ammonia production facilities in Trinidad and Tobago. As another example, CF Industries has production and customers in both the United States and Canada, and a considerable amount of Canadian production is exported to the upper Midwest. In general, the ammonia industry is internationally interconnected for a variety of reasons.

At a basic level, the Haber–Bosch process combines feedstock hydrogen and nitrogen under high temperature (350–500 °C) and pressure (150–300 bar) while passing through a catalyst bed. Ammonia is produced via the following exothermic reaction:

$$3H_2 + N_2 \rightleftharpoons NH_3 \quad \Delta H = -91.8\,\text{kJ/mol} \tag{1.1}$$

This ammonia synthesis reaction generally yields 15 to 20 % ammonia in a single pass over the catalyst bed. The reactor is cooled to below 0 °C, resulting in condensation of ammonia, while hydrogen and nitrogen remain in the gas phase. The unreacted hydrogen and nitrogen are recycled back to the reactor and combined with new feedstock. State-of-the-art implementations of the Haber–Bosch process tend to use multiple catalyst beds in series with intercooling to improve reaction conversion and reduce the amount of gas being recycled, ultimately to mitigate synthesis loop pressure drop and reduce refrigeration requirements for condensation. This is achieved either through indirect cooling with a heat exchange network intertwined with the beds or through direct quench cooling, where low-temperature feed gases are partitioned between each bed's inlet, which also serves to increase synthesis rates by reducing ammonia concentration [184].

At present, hydrogen is largely obtained by reforming natural gas and coal. Nitrogen is obtained from air either through the addition of combustion air during the fossil fuel reforming process or directly separated in an electrically powered process. During the 20th century, the goal while developing the Haber–Bosch process and upstream feedstock production was to achieve the most efficient and cost-effective production plants possible. Therefore, particular attention was paid to the cost of hydrogen production feedstock (e. g., natural gas, coal, propane, or water), the energy intensity of the process, and the capital cost of the facilities [301]. Initially, hydroelectric power was used to produce ammonia on the 100,000 mt/y scale with the generated electricity used in electrolysis to produce hydrogen from water and to produce nitrogen from air [314]. This hydroelectric power was inexpensive, but the capital costs of the generating facilities as well as the electrolysis were high. This original process produced minimal carbon dioxide emissions. However, the availability of inexpensive natural gas drove the nitrogen fertilizer industry to shift and expand to world-scale plants that produce more than 1,000,000 mt/y of ammonia. In this context, the synthetic nitrogen fertilizer production paradigm was established and grew into a mature industry. Currently, synthetic ammonia is essential for our way of life.

However, significant sustainability concerns remain with respect to emissions and resource utilization. Coinciding with the development of the modern nitrogen fertilizer production industry in the latter part of the 20th century and especially beginning in the 1980s, researchers began to see a clear trend of the temperature of the planet increasing over time and began to hypothesize that carbon dioxide emissions may cause a greenhouse effect in the atmosphere. As more studies and reviews were performed, concern grew, and other gases were added to the list of recently termed greenhouse gases. In 1988, the issue was elevated within the United Nations (UN) with the formation of the Intergovernmental Panel on Climate Change (IPCC). The goal of the IPCC was and continues to be to provide policymakers with regular scientific assessments on the current state of knowledge about climate change. For the fourth report in 2007, the IPCC panel members and former US Vice President Al Gore were jointly awarded the Nobel Peace Prize "for their efforts to build up and disseminate greater knowledge about man-made climate change, and to lay the foundations for the measures that are needed to counteract such change" [63]. This fourth report focused on limiting global warming to 2 °C. Subsequent reports focused on inventories and contributors of greenhouse gases, of which one of the main contributors is the production of anhydrous ammonia and nitrogen fertilizers [64, 62]. Approximately 45 % of the hydrogen produced globally from natural gas and coal serves as a precursor to ammonia [2]. More broadly, it is estimated that ammonia production is responsible for 1–2 % each of global energy consumption and carbon dioxide emissions [26, 232].

1.1 The history of fertilizer supply and utilization in Midwest United States: A motivational case study

Global nitrogen fertilizer companies have continued to look for opportunities to exploit inexpensive energy and raw materials throughout the development of the industry. Ultimately, this has led to large supply chains for transferring nitrogen fertilizer product to regions of demand. One of the largest of these regions of demand is in the Midwest US Corn Belt, which is comprised of Illinois, Indiana, Iowa, Michigan, Minnesota, Missouri, Ohio, and Wisconsin. Initially, nitrogen fertilizer was produced primarily within the Gulf Coast states, which are home to inexpensive natural gas. Supply networks were built that connect the Gulf Coast and Midwest US, including ammonia pipelines, barges, and train transport. Since nitrogen fertilizer was applied to crops during limited times of the year, the supply network included vast regional storage facilities, as well as smaller retail outlet storage facilities. All involved in the storage and transport of ammonia became acutely aware of its safety hazards. If it is released, it will burn the skin, lungs, eyes, and other parts of the body. As natural gas prices have become more competitive across the world, ammonia and other forms of nitrogen fertilizers have begun to be delivered from countries with high natural gas reserves into the supply network through Gulf Coast ports for shipment to the Midwest and across the US. During the same period, Canadian nitrogen fertilizer began to flow from the north into the United States. With global competition came industry consolidation, with a few companies controlling most of the production and regional distribution. Smaller farmer-owned cooperatives came together to form separate regional cooperative structures that had more purchasing power and could also buy into the fertilizer industry, thus vertically integrating by owning the production of crop inputs. These larger cooperative structures were initially intended to serve the smaller cooperatives, but in some cases ended up competing against them. Smaller private companies that supply farmers also emerged and then consolidated to form regional, national, and international companies.

As the fertilizer industry was transforming into what it is today, farmers began to understand how to maximize crop production. At the farm level, nurse tanks of liquid ammonia are delivered to a field. A tractor pulls a large tool bar with plastic tubes leading to pipe fitted knives that cut into the soil. The nurse tank is pulled behind and the valves are remotely opened from the tractor delivering liquid ammonia to the knives. As ammonia is exposed to the subsoil, there is a quick shift from liquid to gaseous phase ammonia as a result of the change from a pressurized tank to atmospheric pressure. The ammonia then binds to water to form ammonium within the soil. In contrast, urea is typically spread in farm fields with a pull behind or self-propelled spreader and then worked into the soil with a tillage implement. The urea may also include a mixture of phosphorus, potassium, and micro-nutrients. Dry and liquid nitrogen fertilizer can be applied by other means such as side dressing with a tool bar or integrated as part of the planter (corn) or seeder (small grain). As supply chains matured, nitrogen fertilizer was inexpensive and dramatically increased yields. Recommendations from universities and/or other public-facing research institutions were typically designed to maximize yields, without considering other factors such as environmental impacts. As decades passed, nitrogen fertilizer costs became prohibitively high and research showed the connection between overfertilization and impacts on the environment such as eutrophication. In response to both cost and environmental concerns, fertilizer recommendations were modified to only provide the correct amounts needed to optimize corn and small grain production. There was no longer an incentive to add nitrogen fertilizer beyond what the crop could use. The development of precision agriculture has now provided the opportunity to place the exact amount of fertilizer needed in the exact placement required for the plant. Slow release variations of nitrogen fertilizers have also improved the utilization of nutrients that are provided in the correct stage of plant growth. Currently, farmers in the Midwest US use less fertilizer than they used to, but the nitrogen fertilizer they apply still accounts for 80 % of US-based carbon dioxide emissions from ammonia use. At the same time, these farmers are subject to fluctuating fertilizer prices, which makes their year-over-year economics volatile. In general, it can be argued that current nitrogen fertilizer practices are not broadly sustainable from a life-cycle impact or economic perspective.

1.2 Sustainable ammonia production and its benefits

In this book, we use the term **sustainable ammonia production** to refer to the practice of making ammonia using energy and feedstock sources with considerably lower carbon intensity, that is, emissions per ton of product, relative to the conventional approach of making ammonia using fossil fuels. The main difference between the sustainable ammonia production discussed in this book compared to conventional ammonia production is the approach to producing feedstock hydrogen. Hydrogen production is responsible for the majority of energy consumption and carbon emissions in ammonia production from fossil fuels. Herein, three distinct approaches to sustainable ammonia production are delineated on the basis of their approach to obtaining feedstock hydrogen.

1. Ammonia production using wind and/or solar power to obtain hydrogen via electrolysis
2. Ammonia production using hydroelectric power to obtain hydrogen via electrolysis
3. Ammonia production obtaining hydrogen from biomass and other organic waste

Nitrogen for sustainable ammonia production can be obtained using electrically powered air separation technology. For certain ammonia production pathways, using biomass and organic waste, nitrogen can instead be obtained during the reforming process by adding air in the correct proportions. This occurs when ammonia is produced from hydrogen (and nitrogen) obtained from steam reforming of natural gas. The carbon intensity of hydrogen produced from electrolysis depends almost exclusively on the carbon intensity of the input power, especially because electrolysis is energy intensive. Wind, solar, and hydroelectric power have very low carbon intensity and are considered “renewable” energy resources. This book does not discuss the production of ammonia from natural gas using CO_2 capture to reduce methane emissions, sometimes terms “blue” ammonia, nor does it discuss the production of ammonia produced via electrolysis powered by nuclear energy.

Blue ammonia production has been viewed as a transitional approach to retrofit existing methane-fed ammonia plants and reduce their carbon intensity. A key concern with this approach is the location in which this CO_2 would be sequestered. The total global CO_2 sequestration capacity projected by 2030 is 615 million mt/y [155]. Approximately 75 % of this would need to be used for ammonia alone if current global production was to be achieved through the natural gas plus carbon capture approach. Additionally, climate think tanks have been skeptical of the true reductions in carbon intensity afforded by this approach to ammonia production due to upstream methane emissions [402, 95]. Methane has a global warming potential over 80 times that of CO_2

over a 20-year period. More broadly, these groups have expressed concern that the construction of new ammonia production based on natural gas with carbon capture will further entrench fossil fuels to an unsustainable degree [402, 95]. Multiple existing ammonia production facilities are nonetheless being retrofitted with CO_2 capture in the near term.

Nuclear energy is not likely to be a widespread solution to making ammonia production more sustainable. Many countries are phasing out nuclear power or have altogether made illegal nonmilitary power generation over concerns about waste disposal and the possibility of power plant accident, which can have catastrophic consequences. The construction of new nuclear generation in countries without prohibitions on its use face extensive permitting and cost overrun barriers. In fact, the levelized cost of new nuclear generation has been found to be higher than the levelized cost of intermittent wind or solar generation plus firming that intermittency through energy storage [208]. Overall, the argument can be made that existing nuclear generation, where available, is best served being used to directly reduce the carbon intensity of supplying energy to the power sector.

Sustainable ammonia production provides several potential benefits. First, the carbon intensity of the ammonia product is minimal. This would contribute significantly to an effort toward net zero from the chemical industry. In fact, the decarbonization of ammonia production, together with that of cement, steel, and ethylene, will be unavoidable if net zero emissions are to be achieved [71]. Combating anthropogenic climate change is crucial from a global point of view, but agriculture is subject to some of its worst and most immediate impacts. Weather events are becoming more extreme, ranging from increasing and more severe storms, drought conditions, and wildfires. All of these can have catastrophic outcomes for farms and farmers, while also leading to increased insurance costs, which will likely be passed onto food consumers. Sustainable nitrogen fertilizer production is a so-called "drop-in" replacement. Farmers do not have to change their production practices and equipment lines to significantly and immediately reduce the carbon intensity of agriculture. Regenerative farming practices, which use significantly less fertilizer, are progressing but are not yet ready as a viable alternative to widespread agricultural production, especially with increasing population.

Second, this approach to ammonia production should stabilize the price of ammonia, as wind, solar, or hydroelectric energy is generally contracted for 20- to 30-year time frames through power purchase agreements (PPA). Sustainable ammonia production and clean power generation entities could also theoretically be integrated, whereby the entire production chain would be owned by a single company and power purchases would not need to occur. A stable market could also theoretically be established for biomass feedstocks, especially if crop waste without considerable other value is being used. In contrast, natural gas prices are very volatile, which is reflected in the nitrogen fertilizer market. For example, the price of ammonia in the US Gulf Coast spot market has been as low as \$200/mt and above \$1,000/mt in the last 20 years [288]. After factoring in transportation costs and markups, US farmers have paid more than \$1,600/mt

[306]. This specifically creates a problem for crop farmers because nitrogen fertilizer represents their largest annual expense. A stable ammonia production cost that could translate into stable prices for farmers would certainly reduce the financial pressure created by uncertain input costs.

There is also a potential benefit to a reduction in the ammonia production scale driven by the nature of wind, solar, and biomass resources. Wind and solar generation generally have large spatial footprints and their implementation into world-scale facilities, which could require multiple gigawatts of generation could be limited by land use. Biomass is another inherently distributed resource. It has a low value density at the point of collection and transportation over long distances would likely be cost prohibitive. In both cases, smaller ammonia production facilities distributed closer to these resources and possibly to fertilizer demands could be a solution. This would lead to some loss of economies of scale, but it would also reduce the cost, emissions, and hazards associated with long-distance ammonia transportation. Transportation costs within the United States can be as high as $150/mt [18] and can even exceed $500/mt in other parts of the world [107]. Beyond mitigating transportation costs, distributed ammonia production could ensure the security of fertilizer supply during supply chain disruptions. In contrast, the existing global supply chain can be severely impacted by fossil fuel shortages or instability in fertilizer-producing regions. Finally, distributed production is likely to bring economic benefits to a broader set of stakeholders. For example, replacing a single 1,000,000 mt/y fossil-based facility with twenty 50,000 mt/y sustainable facilities that are optimally distributed relative to input resources and fertilizer demand would result in net job creation and bring revenue to 19 new communities. This phenomenon could be especially important for rural economic development.

In addition to the benefits for climate and economic sustainability mentioned above, all-electric ammonia synthesis powered by wind, solar, or hydroelectric power offers certain technical advantages. Electrolysis can produce hydrogen at higher pressure (more than 30 bar), which reduces the compression requirements of the synthesis loop [347]. Furthermore, obtaining hydrogen from electrolysis and nitrogen from electrical air separation allows for close-to-pure feedstock and requires less purge than a fossil-based feedstock to maintain a 3:1 molar ratio of hydrogen to nitrogen. This improves overall conversion of feedstock and energy efficiency [347].

Other possibilities include further processing of sustainable ammonia into urea, ammonium nitrate, or ammonium sulfate. The production of "sustainable" urea is intriguing, as the Midwest has a large concentration of bioethanol plants that use mainly corn grain as feedstock. Corn grain is fermented, resulting in the release of high-purity biogenic carbon dioxide. This carbon dioxide may be captured and combined with sustainable ammonia to produce "sustainable" urea. Urea has some advantages over ammonia as it is safer to handle, is applied quickly to farm fields, and is less costly and easier to store. Farm cooperatives that provide crop input are numerous throughout the Midwest. Almost every agricultural community has a cooperative that stores ammonia and urea. However, in perspective, there is a much larger amount of urea storage available at cooperatives and, in some cases, on farms.

1.2.1 Sustainable ammonia to enable sustainable energy

The promise of renewable ammonia is not limited to fertilizer sustainability alone. Commercial power generation is another pillar of modern society. Traditionally, fossil fuels have been used to generate electricity in large power plants. However, as with agriculture, sustainability concerns regarding carbon dioxide emissions and resource availability have led to a global push to increase the penetration of renewable energy into the mix of power supply [4]. This transition is already underway. According to current projections, the national penetration of renewables in the US is expected to reach at least 40 % by 2050, most of this increase being due to wind and solar [372]. However, wind and solar are intermittent in that their availability occurs at irregular intervals as a result of time-varying weather phenomena such as wind speed and solar radiation. This means that wind and solar generation are not continuous or steady. As a result, energy storage is needed to balance this intermittent generation with power demands. Moreover, low-cost storage is needed for cost competitiveness because large amounts of energy stored over long periods (e. g., months or seasons) will be needed to accommodate high wind and solar generation [434].

Batteries are the technology most commonly and historically considered for energy storage, but their high cost and relatively low energy density make their use for seasonal storage prohibitively expensive [9, 37, 302]. Electrolytic hydrogen has also been considered as an energy storage medium. Specifically, hydrogen could be produced during periods of excess renewable availability, stored, and later used for power generation during renewable generation deficits, most commonly via fuel cells. This power-hydrogen-power concept is well established and has received considerable research and development [88, 106]. The main limitation of hydrogen-based energy storage is the high cost of the storage itself. Hydrogen is the least dense of all elements, and so volumetric limitations necessitate that it be stored as either a high-pressure gas or a cryogenic liquid. Gaseous hydrogen storage is expensive due to the type and amount of material needed due to high pressure, whereas liquefaction of hydrogen requires a significant amount of energy beyond that needed for the already energy-intensive electrolysis [121, 80].

The further processing of hydrogen into ammonia is a potential route to decrease energy storage costs [193, 415, 204, 273]. Ammonia can be stored as a liquid at ambient temperature and moderate pressures, making it considerably less expensive to store than hydrogen. Liquid ammonia contains three times the energy of gaseous hydrogen or 70 % more energy than liquified hydrogen on a volumetric basis [205]. Figure 1.4 shows that ammonia is best suited for energy storage at seasonal storage durations of at least a month to a year and on a scale of multiple terawatt hours (TWh). This is a clear contrast with hydrogen and even more so batteries. When dispatchable power generation

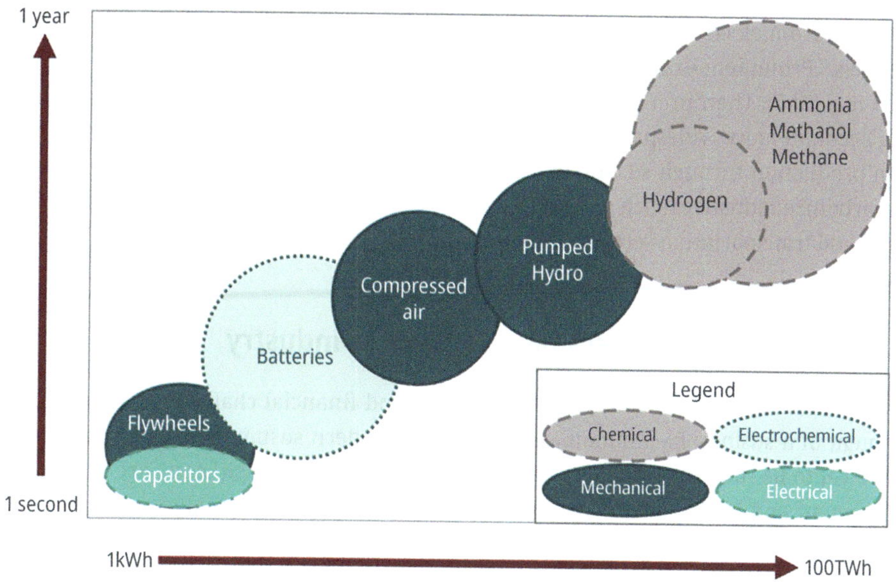

Figure 1.4: A comparison of energy storage technologies and pathways across relevant time scales and capacities. Ammonia is an energy dense storage medium suitable for long-term energy storage.

is needed, hydrogen can be obtained via the endothermic decomposition reaction of ammonia back to hydrogen and nitrogen. This reaction is commonly termed ammonia "cracking." Ammonia can also be used to directly generate electrical power through ammonia-fed fuel cells or traditional fossil fuel technologies such as internal combustion engine generator sets or combustion turbines with modifications to mitigate nitrogen oxide (NOx) emissions and corrosion [410, 194]. Both ammonia cracking catalysts [202] and direct-fed fuel cells [8, 378] are active topics of experimental research. Meanwhile, internal combustion of fossil fuel ammonia mixtures [148], hydrogen ammonia mixtures [260, 103], and pure ammonia [198, 179] is in the demonstration stage. The ability to use ammonia in an internal combustion engine also allows its use as fuel. In an agricultural setting, this could include fuel for tractors [362] or for grain drying in corn production [356]. Ammonia has also been proposed and investigated as a marine fuel [127, 17]. In general, sustainable ammonia for energy applications is a natural extension of the power-to-hydrogen concept with the distinct advantage of lower storage costs.

Another benefit of deploying sustainable ammonia production would be the potential far-reaching implications of infrastructure development beyond the ammonia industry itself. For example, a sustainable ammonia production industry would entail a considerable amount of sustainable hydrogen production. This would certainly drive many aspects of the industry, including electrolysis manufacturing. The scale-up of electrolysis manufacturing could allow for more widespread uses of sustainable hydrogen such as the production of iron and steel with lower carbon intensity as well as fuels synthesized from electrolytic hydrogen and carbon dioxide, which are commonly termed "e-fuels." Prominent examples of these e-fuels include e-methanol and sustainable aviation fuel (SAF). Their production becomes much more viable if electrolysis manufacturing becomes more widespread. In general, establishing a foothold in sustainable hydrogen production through sustainable ammonia is a gateway to decarbonizing hard-to-decarbonize industries such as iron and steel production (8 % of global GHG emissions) [154] and transportation sectors (23 % of global GHG emissions) [170].

1.3 Challenges to a sustainable ammonia industry

There remain several technical, market, policy, and financial challenges to the development of a sustainable ammonia industry. Few modern sustainable ammonia plants have yet to be fully operational in the industrial sectors, although many are under development and in the early stages of commissioning. One of the challenges is that the Haber–Bosch ammonia synthesis process has a long history of operating under steady-state operation. This process would likely require significant changes to its technical design and operational strategy if a variable hydrogen supply is provided to the reactor due to intermittency of wind or solar power. Although this has been proven technically feasible by multiple commercial entities as discussed in Chapter 5, unproven technical designs increase financial risks and potentially the cost of financing. The scale of the electrolysis required for sustainable ammonia production at scale is much larger than that previously observed, even though electrolysis manufacturing has occurred for decades. Questions remain about the ability to quickly scale up electrolysis manufacturing while still ensuring that their performance remains robust and durable. Electrolysis supply chains are still developing, raising questions about components sourced as well as maintenance and service. From a work force standpoint, training and experience are not necessarily readily available in the regions (e. g., the Midwest) where sustainable ammonia industries may be developed.

It can be difficult to compete head-to-head with natural gas-based ammonia production from a market price standpoint. This global market is volatile, but low natural gas prices can lead to low ammonia production costs. Establishing the market price for new sustainable ammonia production is challenging. Questions remain whether this new commodity would need to compete directly with conventional ammonia prices or whether a premium would be paid for sustainable ammonia. One possible approach to mitigate the market barrier is to sell ammonia or other derivatives directly to farmers who consume these fertilizers. In many respects, farmers "own" the demand for nitrogen fertilizer. A local farmer cooperative could develop and own sustainable nitrogen fertilizer production facilities based on ammonia or join a group of cooperatives to develop and operate a larger-scale production facility. The further processing of sustainable ammonia to urea is an approach that could be considered in this case. As explained earlier in the chapter, most ammonia is typically stored in large, regional refrigerated storage facilities. These large storage facilities are typically controlled by a handful of large fertilizer companies. Ammonia is typically delivered from regional storage to retail facilities at the local cooperative level, which are much smaller and cannot handle seasonal or annual storage volumes. Ammonia is only applied in spring, early summer, and fall. A sustainable ammonia production facility would need to produce and store the product throughout the year. As a result, a farmer cooperative that desires to produce their own sustainable nitrogen fertilizer would likely be obligated by these larger fertilizer companies to store ammonia throughout the year until times of need. The further processing of sustainable ammonia to urea using carbon dioxide provides a potential pathway to independence from these larger companies. Ammonia is commonly transported via tanker trucks across the Midwest US where it is directly applied as fertilizer. Sustainable ammonia produced on a wind farm could easily be trucked to carbon dioxide sources, such as ethanol plants, which could have individual urea plants on site. This eliminates the need to pipe, rail, or truck hydrogen or carbon dioxide in the production of sustainable nitrogen fertilizer.

Policies are being implemented at both the federal and state levels within the US to support sustainable ammonia production. Programs through the US Department of Agriculture and the Departments of Agriculture of many US states have provided grants to begin implementation of this new industry. The 2022 US Bipartisan Infrastructure Act provided funding for the establishment of hydrogen hubs across the country. A portion of the hubs that were awarded funding included sustainable ammonia production as part of their hub development plan. The impact of these federal government programs, while valuable, is small in comparison to the potential impact of the US Inflation Reduction Act (IRA), which was also passed in 2022 [6]. Section 45V of the IRA provides a hydrogen production incentive of \$3/kg of hydrogen if production is achieved with carbon intensity below 0.45 $mtCO_2/mtH_2$. This policy has transformative potential to make sustainable ammonia production competitive with conventional ammonia production.

However, the publication of the final guidance for Section 45V of the IRA was delayed until early 2025 due to the complexity of determining specific implementation rules. These rules included specifications around (i) the timescale on which carbon intensity is calculated, (ii) the larger-scale implications (e. g., additionality) of using existing renewable generation to make hydrogen, and (iii) using clean power from a certain region in the United States to make hydrogen in a different region [371]. The carbon intensity calculation timescale and regionality requirements were relatively favorable from the standpoint of new projects. Carbon intensity is to be calculated on an annual average basis up to 2030, at which point this will transition to an hourly average basis. Hydrogen production and renewable generation must occur in the same region as defined by the US Department of Energy 2023 National Transmission Needs Study, of which there are 15 [370]. However, from an additional standpoint, new hydrogen products are required to use renewable generation constructed within 3 years prior of hydrogen production beginning. This altered the plans of certain entities who previously envisioned using already existing generation assets, especially given the long delivery time for electrolysis.

In addition to this barrier, finalizing these rules required cooperation between different federal agencies, which meant that this final guidance was not released for more than two years after the law was passed. These agencies included the US Internal Revenue Service (IRS), which is responsible for administering the tax credits, the Department of Energy (DOE), which has the collective technical expertise to understand the impact of specific credit rules on adoption, economic benefit, and climate impacts, and the Biden administration executive branch, which had its economic development and climate priorities. Furthermore, legislative uncertainty around the rules and funding availability for these credits with changing presidential administrations has slowed the pace of renewable ammonia deployment in the United States.

2 Sustainable ammonia production: History and current progress

This chapter discusses sustainable ammonia production from many different perspectives. Then it summarizes the history of sustainable ammonia production efforts from hydroelectric power, biomass, and wind and solar energy. Today, more than 90 % of the global ammonia production uses fossil fuels, but this dominance was not always the case [314]. In the 1930s, 30 % of global ammonia production was powered by hydroelectric generation.

The chapter then outlines current efforts to produce ammonia sustainably from each of these feedstocks, as well as describing the future outlook of sustainable ammonia production.

2.1 A brief history of sustainable ammonia production

The first sustainable ammonia production process was invented by Casale and Fauser in Italy in 1921 [330]. This process used hydroelectric power to separate hydrogen from water by alkaline electrolysis and nitrogen from air via cyrogenic distillation [314]. At the time, alkaline electrolysis had a best-case efficiency of 57.8 MWh/mtH_2. Today, this efficiency has been reduced to as low as 46.7 MWh/mtH_2. The ammonia synthesis technology coupled with the generation of hydroelectric power was the Casale process, which operated at pressures above 800 bar and did not require refrigeration for liquefaction. In comparison, the Fauser process operated at pressures of 250 to 300 bar, similar to the Haber–Bosch process [364]. The Casale process began operating in Terni, Italy, in 1921 and produced 10,500 mt/y at its peak production in 1927. By the late 1920s, the Montecatini corporation, which employed the Fauser process, monopolized the ammonia industry in Italy and was producing more than 140,000 mt/y. The Casale facility in Terni could not compete in the ammonia market due to economies of scale and instead became primarily a research facility. This research-oriented approach has served Casale well; they have driven considerable innovation in the ammonia industry, especially as it pertains to the renaissance of sustainable ammonia production more than 100 years from its initial implementation.

At the same time, the hydroelectric-powered Casale process began to be implemented in other European countries due to its worldwide license. Before 1930, Casale-based hydroelectric ammonia facilities were operating throughout Europe at relatively small scales, specifically in Saint Auban, France (3,500 mt/y), Sabiñánigo, Spain (5,000 mt/y), Visp, Switzerland (2,800 mt/y), and Dugi Rat, Yugoslavia (3,500 mt/y). Even larger facilities were operating in Nobeoka and Minimata, Japan, which had a cumulative hydroelectric ammonia production capacity greater than 50,000 mt/y by 1927 [92]. Casale also exported their technology to the United States and Canada. A 3,000 mt/y fa-

https://doi.org/10.1515/9783110762525-002

cility operated in Niagara Falls (NY), USA, from 1924 to 1927 while a 2,500 mt/y plant was installed in Niagara Falls (ON), Canada, in 1930. By the 1930s, the Fauser process was also implemented globally, with a facility in Ljunga, Sweden as well as a 32,000 mt/y plant in Trail (BC), Canada. The facility in trail eventually achieved a capacity of 70,000 mt/y [67].

In the mid-1920s, the first attempts at sustainable ammonia production from biomass were also made in the United States. Specifically, these plants were installed in the Corn Belt region of the United States, which dominates national corn production. The hydrogen feedstock for these plants was produced through corn fermentation. Two facilities were installed, the first in Peoria (IL) in 1926 and then in Terre Haute (IN) in 1927. They had respective capacities of 4,500 mt/y and 5,500 mt/y. Ammonia production at these facilities was short-lived. By 1928, both facilities redirected their operations to instead produce methanol. The hydrogen feedstock was considered to be more valuable in the context of methanol production and, as a result, the downstream chemical process equipment was repurposed [365].

The adoption of this process continued to grow through the 1930s, by which time 30 % of global ammonia production came from hydroelectric-powered processes (see Figure 2.1). Many of these ammonia production facilities were located adjacent to existing hydroelectric generation. Much of this power was previously used to produce the contemporary nitrogen fertilizers of the time. Specifically, the Birkeland–Eyde process was used to produce nitrogen oxides (NO_x), which were subsequently mixed with water to form nitric acid (HNO_3) [313] while the Frank–Caro process was used to produce calcium cyanamide ($CaCN_2$) [181]. The electrolysis-based ammonia synthesis process was four times more efficient than the Birkeland–Eyde process (15.9 MWh/mt vs. 61.5 MWh/mt) and less capital intensive than the Frank–Caro process. The scale of new

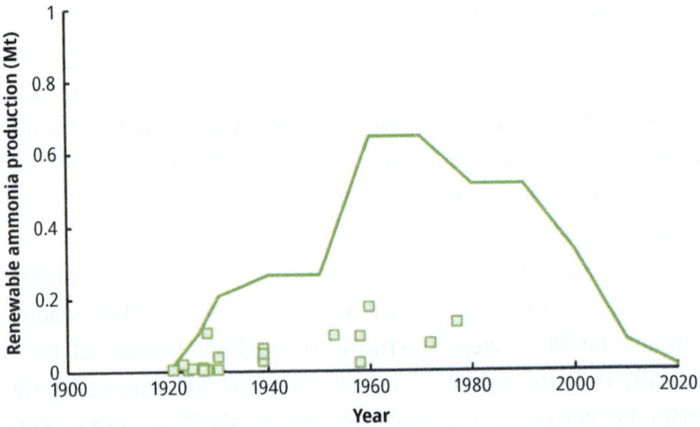

Figure 2.1: Global sustainable ammonia production capacity from 1900 to present. The green line indicates cumulative capacity while green squares indicate individual plants. Reprinted from [314] under Creative Common CC BY License.

hydroelectric ammonia production facilities increased by at least an order of magnitude (>20,000 mt/y) to take advantage of economies of scale; the earlier small facilities could no longer be competitive.

The development of hydroelectric ammonia production led to some early innovation and improvement in electrolysis technology. Alkaline electrolysis systems need to be replaced every 5 to 10 years. This pales in comparison to the multi-decade lifetime for the synthesis plants. This resulted in multiple opportunities for improved successive electrolysis units to be used in the same ammonia facility. There was strong competition among manufacturers, and it was not uncommon for replacement electrolysis units to be supplied by companies other than the original supplier. Overall, this drove an increase in the electrolysis efficiency, and thus the efficiency of the entire hydroelectric ammonia synthesis process. In Rujkan, Norway, a 125,000 mt/y hydroelectric-powered facility based on the Haber–Bosch process became operational in 1927 [115]. This facility used generation that was previously allocated to the production of nitric acid. The synthesis loop was provided by the predecessor to BASF while electrolysis was provided by Norsk Hydro. Both of these companies are among the leaders in their respective industries today, almost 100 years later. Another hydroelectric-powered facility using Norsk Hydro electrolysis began operations in Glomfjord, Norway, in 1953. This facility had a nameplate capacity of 110,000 mt/y [115].

In 1958, a 100,000 mt/y plant was built in Nangal, India, sourcing hydroelectric power from the Bhakra Dam [35]. Electrolysis was supplied by the Italian company De Nora, which still exists today. The ammonia made here was further processed to 400,000 mt/y calcium ammonium nitrate. The first sustainable ammonia production facility in Africa was built in Aswan, Egypt, in 1960. The motivation for building this facility was to enable national food security through the production of domestic nitrogen fertilizers. This plant had an annual capacity of 175,000 mt/y and was powered by hydroelectric generation from the Aswan Dam. Electrolysis at plant start-up was supplied by De Nora, with a later replacement coming from Brown Boveri, which is now part of ABB. To date, this is the largest sustainable ammonia plant ever built. Hydroelectric ammonia production also spread to South America in 1962. A 20,000 mt/y facility was built in Cusco, Peru. This plant used eight 3.5 MW high pressure (30 bar) electrolysis modules provided by Lurgi. A facility using this same electrolysis technology was built in Kwekwe, Zimbabwe, in 1975. This plant used 28 electrolysis modules powered by the Kariba Dam for a total ammonia production capacity of 72,600 mt/y. As of this writing, this is the last sustainable ammonia plant built with a capacity above 10,000 mt/y.

The peak of sustainable ammonia was from 1960 to 1970 with a global production capacity of approximately 650,000 mt/y [314]. From this point forward, there was no new development of sustainable production with the exception of the aforementioned Zimbabwe facility. Many of the plants described above were decommissioned. Some facilities in Italy and Japan were destroyed by bombing during World War II. Other facilities were commandeered or stripped for parts by Allied forces in Italy and Russian Red Army forces in Japan after the war. Beyond this, there were several reasons for the decommissioning of sustainable ammonia plants elsewhere.

In the 1940s and 1950s, a demand for hydrogen in the petrochemical sector led to the advent of natural gas reforming processes [47]. Natural gas contains less sulfur than coal and, as such, requires fewer unit operations and consumes less energy in hydrogen production. Both natural gas- and coal-based ammonia production also benefited from downstream heat integration to improve energy efficiency as well as economies of scale. Process optimization enabled integration of upstream reforming and downstream synthesis to reduce energy consumption for ammonia production from natural gas and coal to 30 GJ/mt and 45 GJ/mt by 1975 (see Figure 2.2). Even in the 1920s, coal was less expensive than hydroelectric generation as a power source. However, obtaining hydrogen from electrolysis was 50 % more energy efficient than obtaining it from coal. However, reforming technology improvements has brought the energy efficiency of coal- and electrolysis-based hydrogen production close to parity. This led coal-based ammonia production to be less expensive. Meanwhile, fossil-based ammonia production facilities began having nameplate capacities greater than 350,000 mt/y to greater than 600,000 mt/y by 1980. This is at least double the capacity of the largest 175,000 mt/y facility in Aswan, Egypt.

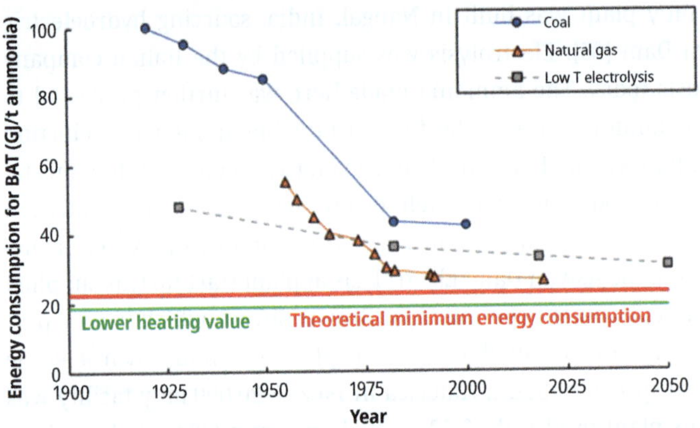

Figure 2.2: Efficiency of ammonia production from 1900 projected to 2050 using hydrogen sourced from natural gas, coal, and low temperature electrolysis. Reprinted from [314] under Creative Common CC BY License.

During the same time, fossil feedstocks, especially natural gas, became more abundant and less expensive around the world. By 1975, state-of-the-art natural gas ammonia production used less than 30 GJ/mt [377]. This corresponds to an ammonia feedstock cost of approximately \$30/mt per \$/GJ. In other words, natural gas that costs \$2.07/GJ, the US Henry Hub global average price in the second quarter of 2024 [374], would result in an ammonia feedstock cost of \$62.1/mt. In contrast, hydroelectric ammonia production in

1975 required at least 10.5 MWh/mt. An optimistic hydroelectric power cost of $30/MWh would thus result in an ammonia feedstock cost of at least $315/mt. This more than fourfold feedstock cost became difficult to deal with. This was exacerbated as the nitrogen fertilizer industry became globalized and ammonia could be manufactured in locations with cheap and easy access to natural gas.

The Rjukan, Norway plant operated for 44 years until 1971 while the Glomfjord, Norway, facility operated for 38 years until 1991. This is in line with the range of lifetimes for an ammonia plant, although at the lower end. By this time, the relatively low cost of natural gas and the technological improvement of the associated technologies made the continuous maintenance of the upgrades of a hydroelectric ammonia plant uneconomical. Both of these plants were decommissioned. The Rjukan hydroelectric generation was also decommissioned in 1971, while the Glomfjord hydroelectric generation still provides power to the Norwegian grid today.

The feedstock for the Indian facilities was switched to fossil fuels because of economics. In 1978, electrolysis at the India facility began to operate at decreasing partial capacities, while hydrogen for ammonia production was obtained from the reforming of the fuel oil [59]. By 2013, this plant had switched to natural gas as a hydrogen source [248]. CO_2 byproduct of these fossil feedstocks is currently used to produce urea from this ammonia. A new natural gas-fed ammonia plant in Egypt was built in Aswan, near the existing hydroelectric-powered ammonia plant, in the mid-2010s [24]. This facility ultimately produces urea. The hydroelectric-powered facility was decommissioned in 2019, after 59 years of operation.

Broader energy considerations have also led to the repurposing of hydroelectric generation away from ammonia production. The Trail (BC), a Canadian plant, operated until the American energy crisis in the 1970s. After this, exporting Canadian hydropower to the US was allowed, and this became more profitable than using this energy to make ammonia. Thus, the upstream electrolysis section of this plant was abandoned. Natural gas was instead used as feedstock to allow the plant to continue making ammonia. Similarly, the Kweke plant in Zimbabwe operated until 2015. At this point, a national energy crisis led to high electricity prices and this hydroelectric power was repurposed to the power sector. Ammonia production at the Kweke plant was abandoned as a result.

Today, only the Cusco, Peru, plant is operational. It now uses ThyssenKrupp electrolysis, which produces hydrogen at atmospheric pressure. These units are more efficient than the original Lurgi units, allowing a cumulative 24.5 MW capacity from seven 3.5 MW modules to be replaced by only two 10 MW modules. The overall efficiency of this facility is as low as 9.45 MWh/mt. The plant is operated by Enaex, whose main use of ammonia is the production of ammonium nitrate explosives [89]. Despite the relative lack of sustainable ammonia production at scale in the present, the legacy of its 100 years of history is still strong. Many of the companies at the forefront of its initial implementation have remained in place until today; some are leaders in the conventional ammonia industry such as Casale and BASF, and others will certainly play a strong role in the growing sustainable ammonia industry likely driven by electrolysis.

2.2 A resurgence of sustainable ammonia production since 2010

The 2010s saw a resurgence of efforts in sustainable ammonia production, especially focusing on using wind and solar resources to do so. Table 2.1 outlines the currently operational sustainable ammonia production facilities installed since 2013. The projects developed between 2013 and 2018 had very small production scales below 50 mt/y; the purpose of the pilot projects was proof of this concept. These early projects often had public–private partnerships, whether with government, university research institutions, or both. The goal of these pilot-scale demonstrations was not profitability, but rather to test different technology components and operating strategies while also showing that ammonia can be used not only as fertilizer but also as a fuel for tractors or power generation. These pilot projects coincided largely with the global momentum behind sustainable ammonia production, which led to a significant increase in academic research, corporate research and development, and government policy. By 2024, small commercial installations of sustainable ammonia production began to be built. Commercial production efforts came from established fertilizer producers Topsoe and Yara, as well as a start-up company, Talus Renewables. Currently, these small-scale commercial operations sell sustainable ammonia as fertilizer. Their value proposition relies on various factors to be competitive, such as high ammonia transportation costs, a volatile conventional ammonia market, or demand-side pressure to reduce carbon intensity.

Table 2.1: Location, scale, energy source, and installation year for existing sustainable ammonia production projects.

Location	Energy Source	Production Scale (mt/y)	Electrolysis Scale (MW)	Year
Morris (MN), USA [308]	Wind	25	0.041	2013
Blairstown (IA), USA [329]	PV	5	0.007	2015
Oxford, UK [73]	Wind	10	0.013	2018
Fukishima, Japan [176]	Wind and PV	11	0.016	2018
Navaisha, Kenya [357]	PV	350	1.25	2024
Boone (IA), USA [113]	PV	7,000	25	2024
Northwest Jutland, Denmark [104]	Wind and PV	5,000	10	2024
Porsgrunn, Norway [266]	Hydropower	20,000	24	2024

Ammonia production from wind energy was first demonstrated at the University of Minnesota West Central Research and Outreach Center's (WCROC) Renewable Hydrogen and Ammonia Pilot Plant in Morris, Minnesota, USA. At the time, utilities did not want to purchase wind energy because production did not match demand very well. Much of the wind energy is produced at night, while there is a lower demand for power as most people sleep. In addition, there was not enough transmission capacity in the system, so the main wind energy sites were considered "stranded." In the green ammonia scenario, wind energy could be monetized by making ammonia from electrolysis-derived hydrogen. Farmers could use green ammonia as a drop-in replacement for identical conventional nitrogen fertilizer to significantly reduce the carbon intensity of their crops. In addition, utilities could potentially accept power when needed and when transmission capacity was available. The original pilot plant was first envisioned in 2002 and the facility was commissioned in 2013. This pilot plant involved steady-state operation using power both from a colocated 1.65 MW wind turbine as well as power purchased from the neighboring utility. Hydrogen is produced by a 41 kW PEM electrolysis unit and nitrogen is produced by a 3 kgN_2/h PSA unit. Both technologies were provided by Proton Onsite, now a subsidiary of NEL. Ammonia synthesis occurred using a scaled-down, purpose-built version of the traditional Haber–Bosch process (see Figure 2.3). This was designed as a proof-of-concept with 65 kg/d production. The energy intensity of this production is greater than 16 $kWh/kgNH_3$ [308]. This is much higher than the state-of-the-art due to a higher energy intensity for the installed electrolysis (>75 kWh/kgH_2) as well as heat loss from the ammonia synthesis loop, among other factors. In 2019, ammonia produced at this pilot plant was also used to demonstrate ammonia-fueled tractor operation at the WCROC farm. A second generation plant is currently in development that will produce 1 mt/day. This new 1 mt/day plant will integrate state-of-the-art ammonia synthesis catalyst and control approaches to demonstrate flexible operation to accommodate variable renewable energy and to exploit low-cost energy in markets with time-varying prices.

Figure 2.3: Scaled-down, purpose-built Haber–Bosch loop installed at wind-to-ammonia pilot plant in Morris, Minnesota, USA. Reprinted from [66] under Creative Common CC BY License.

A similar small-scale ammonia production project based on solar energy was developed in Iowa in 2016. This project, entitled "The Raphael Schmuecker Memorial Solar Hydrogen System" was a private venture installed at Pinehurst Farm in Blairstown, Iowa, USA. [329, 363]. The project used 8.1 kW of solar generation and 7 kW of PEM electrolysis. As with the Morris, MN pilot plant discussed in the previous paragraph, electrolysis was provided by Proton Onsite. A batch ammonia production system was purpose built for this project. Hydrogen and nitrogen are charged from pressurized storage into the ammonia synthesis process in an amount that allows 38 liters of ammonia to be produced per batch (see Figure 2.4). Its production rate is variable depending on the availability of energy, but on average, this system produces 35 kg of ammonia every 2 weeks. The energy intensity of this facility was more than 20 kWh/kgNH_3, with particularly high energy demands for nitrogen and ammonia subsystems. Specifically, the breakdown of the energy consumption for this facility is 64 % for hydrogen, 23 % for nitrogen, and 13 % for ammonia and the balance of the plant components. In comparison, hydrogen production through electrolysis is responsible for more 90 % of energy production in the state-of-the-art ammonia production from electrical power. Despite the farm location of this pilot plant, the ammonia produced on site was used along with additional hydrogen to power a tractor. This was one of the first field demonstrations of ammonia as a fuel.

Figure 2.4: Compressed hydrogen and nitrogen storage cylinders, which feed batch ammonia synthesis process at solar-to-ammonia pilot plant in Blairstown, IA, USA. Reprinted from [329] under Creative Common CC BY License.

This small-scale renewable ammonia production concept made its way to the United Kingdom. Siemens led the development of a $2 million pilot plant at Rutherford Appleton Laboratory near Oxford, UK [73]. This effort received considerable funding from Innovate UK, a federal government agency. It was a joint venture between Siemens, the University of Oxford, Cardiff University, and the UK Science and Technology Facilities Council (STFC). This plant started operating in 2018 and was powered by wind energy.

Siemens supplied a 13 kW PEM electrolysis unit for this plant. The ammonia synthesis loop was custom designed to produce an average of 30 kg/day. The main purpose of this pilot was to demonstrate the potential of ammonia as an energy storage medium. For this reason, a 30 kW engine generator was also installed to use ammonia as fuel to generate power. This was the first pilot project to demonstrate the full path from renewable energy to ammonia and back to power at the same facility.

This renewable ammonia concept also expanded to Japan in a similar time frame. An ammonia manufacturing pilot plant was installed at the Fukishima Renewable Energy Institute (FREA) of the Advanced Industrial Science and Technology (AIST) in 2018 [176]. This plant was built by JCG Holdings Corporation. The development of this pilot plant coincided with research into new catalysts for lower-pressure and lower-temperature synthesis. This plant uses an 11 kW alkaline electrolysis unit powered by a fraction of the respective 300 kW and 500 kW of wind and solar PV installed at FREA. The ammonia synthesis loop has a capacity of 33 kg/day. It operates isothermally, using an oil-based temperature control system. This deviates from commercial-scale ammonia synthesis, which is exothermic because higher surface area-to-volume ratios for reaction vessels at small-scale means that heat losses need to be considered. This ammonia production system has been used to demonstrate some time-varying ("load following") operation in response to "simulated" renewable variability profiles [105]. A second phase pilot plant, which will have a capacity of 4 mt/d, is planned to begin operation in the late 2024 through 2026 [65]. As with the facility in Oxford described in the previous paragraph, this facility also demonstrated power generation using ammonia as a fuel. This was achieved using a 47 kW gas turbine fed with pure ammonia.

In late 2023, Talus Renewables installed a modular PV-powered ammonia production system near Naivasha, Kenya [357]. This is a commercial system: Talus agreed to a 15-year take-off agreement with the Kenya Nut Company to provide fertilizer for its Morendat farm in Naivasha, as well as other operations. The ammonia production module uses 2.1 MW of PV generation and 1.25 MW of alkaline electrolysis to produce an average of 1 mt/d ammonia. It is completely islanded from any type of utility or power grid, so it also uses a battery to keep the facility running at minimum demand overnight. This was the first islanded, fully renewable commercial ammonia production installation since at least 2010. The economics of this concept are more favorable in many parts of Africa than in other parts of the world. Farmers in African countries often pay more than 100 % of the market price in ammonia transportation costs because this ammonia is often produced in other countries or even continents. As a result, renewable ammonia with a higher-than-market price can still be cost competitive. TalusAg also installed its first commercial module in the United States in 2024 at the Landus Agricultural Cooperative in Boone (IA), USA [113]. This facility can produce on average 20 mt/d and uses a combination of PV generation and power purchased from a utility. The economics of this scaled-down modular ammonia production approach rely on the 45V hydrogen production tax credits for competitiveness as well as its ability to produce ammonia at a stable cost, in contrast to the highly volatile conventional ammonia market to which farmers are subject.

In 2024, a project developed by a joint venture between Topsoe, Skovgaard Energy, NEL, and Vestas began producing ammonia in Northwest Jutland, Denmark [104]. This facility has an annual ammonia production capacity of 5,000 mt/y. It uses 50 MW of existing solar power as well as 12 MW of newly installed wind turbines from Vestas. This facility demonstrates the ability to operate flexibly to directly use this renewable energy. This is facilitated by Topsoe's flexible ammonia synthesis technology coupled with 10 MW of PEM electrolysis provided by NEL. This 100 % renewable ammonia production will help avoid approximately 8,200 mtCO_2/year, which would have been produced via conventional production. This project was funded in part by the Danish Energy Technology Development and Demonstration Program (EUDP).

Yara opened the largest sustainable ammonia production facility in 2024 in Porsgrunn, Norway, at Heroya Industrial Park [266]. This involved a partial upstream retrofit of an existing ammonia production facility. Existing hydroelectric power generation is used to power a newly installed 24 MW PEM electrolysis provided by Linde. The resulting clean hydrogen is fed to the existing ammonia synthesis loop to offset some of the higher carbon intensity hydrogen. This electrolysis-derived hydrogen allows for the production of 20,000 mt/y of sustainable ammonia. This ammonia has already been sold to customers in Sweden. Taking a longer view, Yara plans to electrify more than 800,000 mt/y of existing ammonia production at this Heroya plant by 2030. This electrification using existing hydroelectric generation is easier from a technical perspective than using variable wind and solar generation. The close-to-steady nature of hydroelectric generation means that the existing ammonia production facility can continue to operate as designed.

2.3 Sustainable ammonia production to 2030

Broader commercialization of sustainable ammonia production is expected to occur by 2030. This has been incentivized by an emerging market demand for sustainable ammonia as fertilizer as well as by government policy to promote the expansion of this industry. Multiple countries, notably Australia and Chile, have developed versions of a national hydrogen strategy to aid align the investment of both domestic and international companies to fund world-scale projects. Many of these projects are occurring in multiple discrete phases to enable more gradual investment and capacity expansion. Some of these projects are also leveraging existing ammonia production facilities and only replacing upstream hydrogen production with clean power generation and electrolysis. It should be noted at the time of writing that none of the planned projects involve a biomass-to-ammonia pathway. Some possible reasons for this are discussed in Section 2.4.3 of this chapter. The rest of this section is dedicated to discussion of projects, which have been announced in the last 5 years and are planned to begin some aspect of operation by 2030.

CF Industries completed the installation of 20 MW of alkaline electrolysis in late 2024 at its Donaldsonville Complex in Louisiana, United States [102]. Electrolysis was supplied by ThyssenKrupp AG. It will operate close to steady state, enabled by purchasing power from an adjacent utility as well as purchasing renewable energy certificates (RECs) to certify this as clean power. This hydrogen is fed to the existing ammonia production facility, which has a cumulative ammonia capacity of nearly 8,000,000 mt/y and is capable of producing approximately 20,000 mt/y of ammonia. An offtake agreement for this ammonia was signed by POET LLC, a biofuel and sustainable bioproducts producer. This allows POET to reduce the carbon intensity of their bioethanol production by 10 %. This is undoubtedly a demonstration project, but is emblematic of demand-side market forces that create an opportunity for sustainable ammonia production.

Fortescue Energy and Incitec Pivot Limited are in the scoping phase of electrifying ammonia production at Gibson Island in Brisbane, Australia [101]. This project will rely on the installation of 550 MW of electrolysis to be powered by wind and solar energy purchased through power purchase agreements with third-party generation companies. This will facilitate the production of approximately 400,000 mt/y of ammonia. Specifically, ammonia production will depend on the restart of an existing facility in Gibson Island that used to be fed with hydrogen from natural gas but stopped operation in 2021. One notable feature of this project is Fortescue's commitment to using purified recycled water for a portion of the electrolysis feedstock. Water intensity is an important consideration within the framework of sustainable ammonia production. Commercial production was originally scheduled to begin in 2026, but this could be delayed if the scoping and approval process takes longer than originally anticipated. Furthermore, changing market conditions, including increased electricity costs, could further delay or materially change the scope of the project altogether.

Another major sustainable ammonia project that is scheduled to begin production in 2026 is the NEOM Green Hydrogen Complex. This is led by NEOM Green Hydrogen Corporation, a multinational effort led by ACWA Power, Air Products, and NEOM [250, 275]. As of 2023, this project had a cost of $8.4 billion USD and its financing involved 23 banks and other financial institutions. This project will require the construction of 2.2 GW of new solar generation and 1.37 GW of new wind generation along with 400 MW of battery energy storage and a new 190 km power transmission system. Feedstock hydrogen will be produced through 1.1 GW of electrolysis. This project also involves the construction of a new world-scale ammonia production facility, which will have a 1,100,000 mt/y nameplate capacity. The ammonia is already earmarked for export via a 30-year offtake agreement signed by Air Products. This ammonia will act as a hydrogen carrier, whereby hydrogen will be obtained via ammonia decomposition near its desired point of use. As of late 2024, construction on the NEOM project is 60 % complete, and production is still scheduled to begin in mid to late 2026 [11].

AM Green is in the process of converting two existing ammonia facilities to sustainable ammonia production with a target date of 2027 [151]. This involves installation of upstream electrolysis to produce hydrogen from "around-the-clock" clean power obtained by combining hydroelectric, wind, and solar generation. Electrolysis will be installed in two phases, each consisting of approximately 650 MW of advanced pressurized alkaline systems [173]. The existing ammonia synthesis technologies will be retrofitted to be more amenable to electrolytic hydrogen. The first phase of this project targets approximately 500,000 mt/y sustainable ammonia with plans to expand to 1,000,000 mt/y by 2030. The effort involves an international consortium of established companies such as John Cockerill, the electrolysis provider, and Casale, the ammonia production technologies engineering partner. On the offtake side, Yara Green Ammonia and AM Green recently signed an agreement that will result in Yara exporting up to 50 % of new sustainable ammonia produced by 2027.

Mejillones Ammonia Energy also has plans to begin sustainable ammonia production in 2027 at its Volta Project in Mejillones, Chile [252]. It will begin by producing 300,000 mt/y ammonia before ramping up production to 600,000 mt/y by 2030. This project will use 600 MW of solar generation. This generation is located in the Atacama Desert, to the west of Mejillones, which has some of the best solar radiation on the planet. The power will be transmitted to Mejillones, a port area, which allows water to be obtained from seawater desalination and enables the potential export of ammonia to international customers. The total investment in the project is projected to be $2.5 billion USD.

Other major efforts for sustainable ammonia plants on a global scale have production dates approaching 2030. One such project is led by AES Chile, which will produce approximately 700,000 mt/y of sustainable ammonia in Antofagasta, Chile [309]. This project will entail the construction of considerably new renewable power generation and storage, specifically three new wind farms, three new solar farms, a battery energy storage facility, and power transmission to bring this power to the hydrogen and ammonia production site. This project will require approximately 1.7 GW of electrolysis and will produce a total of 217,000 mt/y of clean hydrogen. Approximately 55 % of this will be used to produce ammonia. This ammonia will be used both domestically as fertilizer and to make explosives for the mining industry while also being exported, potentially to South Korea through an agreement with Samsung C&T. The total project cost is approximately $10 billion USD. This project is currently being assessed for environmental impact by the Chilean government [7].

Another project in Chile, this one led by HNH Energy, submitted an environmental impact assessment in 2024 [137]. This project has plans to use 3 GW of wind powered electrolysis to produce 1,300,000 mt/y of sustainable ammonia in the Magallanes region. This project will also obtain electrolysis feedwater through newly constructed seawater desalination so as not to put stress on Chile's freshwater resources. This project will be developed in phases, with 1.4 GW of wind and seawater desalination installed by 2027 and an additional 2.1 GW of wind along with hydrogen and ammonia production installed by

2030. The total project cost is currently estimated at $11 billion USD. This project involves considerable international participation including Copenhagen Infrastructure Partners, AustriaEnergy, and OKO Wind.

The Australian Renewable Energy Hub project, which will be operated by BP, is taking a similar staged approach [25]. This project is based in Pilbara, Western Australia. The first phase of the project will involve development of 1 GW of renewable generation by 2029 to provide power to domestic customers in Port Hedland and Newman, as well as the possibility of electrolysis-based hydrogen production for domestic use. The second phase of the project, expected to be completed in the early 2030s, includes additional development of wind and solar capacity along with the construction of a desalination plant in Port Hedland and a pipeline to bring water to a larger-scale hydrogen production site in Pilbara. This hydrogen will be used by the Australian industry and could necessitate the construction of a hydrogen pipeline depending on where these customers are located. One such use case is to reduce the carbon intensity of iron ore production. By the mid-2030s, the third phase of the project involves renewable generation build-out up to 26 GW. It will also involve the construction of an ammonia production facility between the coastal cities of Port Hedland and Broom to facilitate ammonia export to international customers along with the possibility of domestic ammonia users. The capacity of the ammonia production facility has not been finalized at the time of writing, but could be as high as 9,000,000 mt/y [31].

2.4 Sustainable ammonia production future outlook

This section outlines the future potential of sustainable ammonia production from biomass, hydropower, and variable wind and solar resources. This potential depends on the location and availability of the feedstocks, the existing infrastructure, and the available technologies. In a general sense, producing 200 million mt/y of ammonia will require an energy input of the order of 2000 terawatt hours (TWh) or 7.2 million terajoules (TJ) assuming an energy intensity of 10 MWh/mt. For wind and solar resources, and hydroelectric generation to lesser extent, this will require proportionally higher capacity additions because these forms of generation do not produce around the clock. This section also outlines the challenges and barriers to more widespread adoption. These challenges are multifaceted in nature, with some being specific to the type of feedstock used for ammonia production and others being commonly shared across these different input resources. Two crucial challenges that are highlighted throughout are (i) the need to scale up electrolysis manufacturing in order to make enough hydrogen for ammonia production via wind, solar, and hydroelectric pathways and (ii) the need for more economical scaled-down ammonia production to better match the smaller scale, more distributed nature of wind and solar generation, as well as biomass availability.

2.4.1 Ammonia production from wind and solar power

At present, global wind and solar generation accounts, respectively, for 2,325 TWh and 1,640 TWh in 2023 (see Figures 2.5 and 2.6). For wind, this energy production required 1,020 GW of installed generation, which implies an global average capacity factor of 26 % [159]. A theoretical 100 MW wind farm operating at this capacity factor would produce an average of 26 MW over a typical year. For solar, 1,600 GW of generation were installed by the end of 2023, which implies operation at an average capacity factor 12 % [158]. Making 200 million mt/y ammonia using wind and solar resources would require approximately 50 % of existing combined generation. On an individual generation facility basis, capacity factors can vary significantly by geography. Therefore, sustainable ammonia developers could locate their production in specific locations with better wind or solar potential to reduce the installed generation capacity needed to meet energy demands. For example, there are many locations in the world that can achieve wind capacity factors above 40 % and solar capacity factors over 20 %.

Looking to 2030, generation from PV resources is expected to approximately triple from current levels and could account for up to 16 % of the total primary energy generation [158]. Meanwhile, wind generation should almost double and account for 14 % of total energy production globally [159]. The rapid expansion of wind and solar development is encouraging for the outlook for making ammonia from these resources.

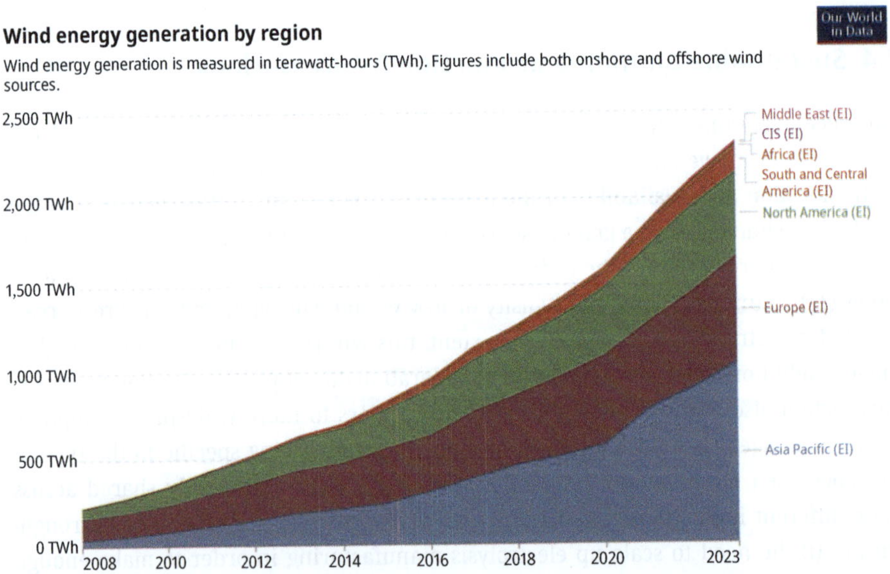

Figure 2.5: Electricity generation from wind by region. This figure is based on gross primary wind generation in each region and does not account for cross-border electricity supply. From Energy Institute – Statistical Review of World Energy (2024) – with major processing by Our World in Data under Creative Common CC BY License.

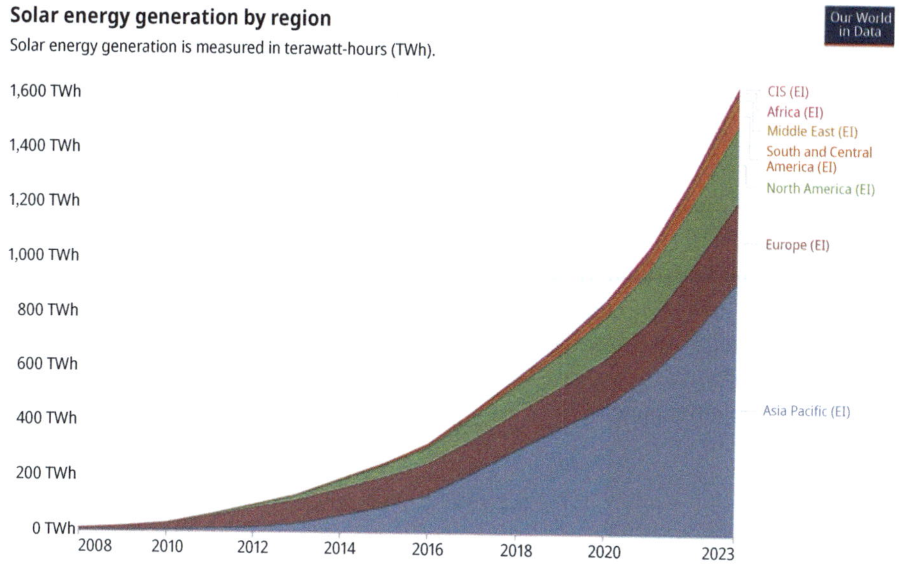

Figure 2.6: Electricity generation from solar by region. This figure is based on gross primary solar generation in each region and does not account for cross-border electricity supply. The data in this figure is from Energy Institute – Statistical Review of World Energy (2024) with major processing by Our World in Data. Reprinted under Creative Common CC BY License.

Making 360 million mt/y of ammonia by 2040 would require approximately 3,600 TWh of energy. This could be sourced from 1,025 GW of wind generation with a 40 % capacity factor, 2,050 GW of PV generation with a 20 % capacity factor, or, in reality, a combination of wind and solar resources, even at the same ammonia production facility.

The ammonia industry and industrial electrification in general will face competition from other sectors for this energy, especially as utilities move to higher proportions of renewables in their energy mix, heating becomes electrified, and the use of battery-electric vehicles expands. One opportunity for ammonia production from wind and solar is the concept of off-grid ("islanded" or "behind-the-meter") facilities. In many parts of the world, especially in the United States and Europe, the pace of transmission expansion lags behind that of adding wind and solar capacity [112]. This results in less than full use of these resources for any applications that rely on the electrical grid to receive power. In order to circumvent transmission limitations, ammonia production facilities could be located close to and directly connect to wind and/or solar generation facilities. This would limit competition for this power to only behind-the-meter applications while also allowing wind and solar developers to more quickly fully monetize their generations assets.

Electrolysis manufacturing must also continue to be scaled up to provide enough capacity for ammonia production along with all other planned industrial uses for hydrogen. At the end of 2023, only 1.4 GW of electrolysis have been deployed [156]. For wind-based ammonia production in a location with the wind capacity factor 40 % this would allow the production of at most 560,000 mt/y ammonia. For solar generation with 20 % capacity factor, this upper limit of ammonia production is closer to 280,000 mt/y. Evidently, orders of magnitude increases in electrolysis deployment are needed. The momentum behind clean hydrogen as an enabler for decarbonization in hard-to-electrify sectors has led to favorable projections for electrolysis manufacturing. The International Energy Agency predicts that by 2030, the global electrolysis manufacturing capacity will reach approximately 400 GW per year according to the announced projects (see Figure 2.7). Overall, producing approximately 360 million mt/y ammonia by 2040 using wind and solar resources will require more than 1,000 GW of electrolysis and potentially up to 2,000 GW depending on the mix of wind and solar used and the specific locations for this generation.

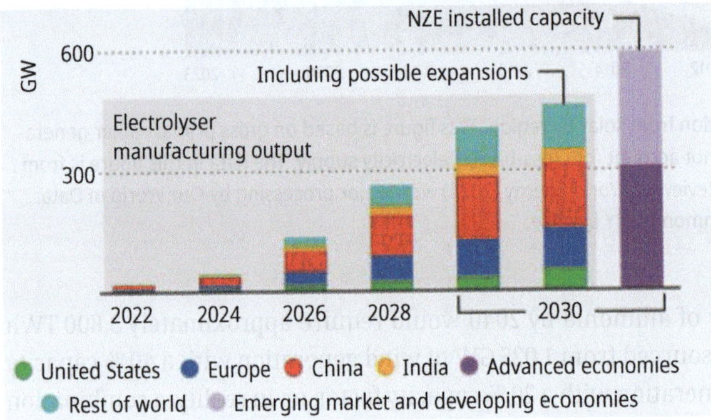

Figure 2.7: Announced cumulative electrolyzer manufacturing capacity output by region. If fully realized, this manufacturing would meet 80 % of the 2030 goal for the IEA 2050 Net Zero Emissions roadmap. Reprinted from [157] under Creative Common CC BY License.

At a macroscopic level, the projected expansion of wind and solar deployment as well as electrolysis manufacturing is promising, albeit somewhat uncertain. However, the scale of production is a critical challenge for individual facilities making ammonia using these resources. Typical ammonia production scales using fossil fuels are more than 1 million mt/y. Achieving this using wind and/or solar resources would likely require between 2.5 and 5 GW of new generation as well as at least 2 GW of electrolysis. Currently, there are less than 10 wind farms in the world with a capacity greater than 2 GW and fewer than 20 solar generation facilities with this capacity. From an electrolysis standpoint, this may also be difficult to achieve in the next 15 years, as 2 GW could be the annual capacity of an entire manufacturing facility. In the near term, it is likely that at least some commercial-scale wind and/or solar-based ammonia production will have production scales of 100,000 mt/y or less, especially for behind-the-meter projects paired with a single wind and/or solar PV farm. These smaller facilities may need innovative approaches to reduce capital intensity and ensure that this deployment is economical. Traditional Haber–Bosch relies on economies of scale, and over the last decades, most of its design decisions are made to maximize energy efficiency. However, the relative importance of capital cost is higher at smaller scales. Innovation in catalysis and separation as discussed in 3 and 4 are promising approaches motivated by this scaling-down consideration.

2.4.2 Ammonia production from hydroelectric power

Globally, there are on the order of 3,700 hydroelectric dams that have capacity greater than 1 MW for a cumulative capacity of 1,400 GW. Hydropower is currently the largest source of clean power generation, producing an approximate annual 4,250 TWh globally (see Figure 2.8). On an order-of-magnitude basis, it would require just under 50 % of this generation to provide power for 200 million mt/y of present-day ammonia production. In 2040, up to 85 % of current hydroelectric generation could be needed to meet projected ammonia demands. However, in reality, this generation is mostly committed to the utility sector in the countries in which it is installed. In fact, hydroelectric generation is often concentrated in certain regions and is crucial to providing the majority

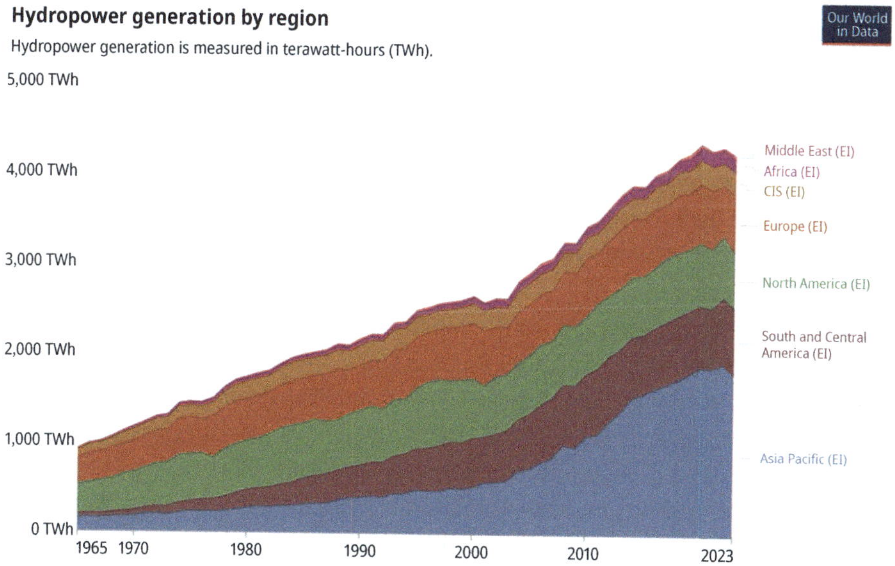

Figure 2.8: Electricity generation from hydropower by region. This figure is based on gross primary hydroelectric generation in each region and does not account for cross-border electricity supply. The data in this figure is from Energy Institute – Statistical Review of World Energy (2024) with major processing by Our World in Data. Reprinted under Creative Common CC BY License.

of power in certain countries or regions of countries. Furthermore, the capacity for hydroelectric generation is not expanding at nearly the same rate as wind and solar generation. Between 2018 and 2023, the global hydroelectric generation capacity increased by less than 1 % per year [161]. Most of these capacity additions occurred in China and India. This may provide some synergy with domestic fertilizer demands, as these two countries have the highest populations in the world.

Much of the stagnation in hydroelectric development is due to long permitting processes and other policies, which make the construction of new hydroelectric dams cumbersome. In addition, these projects often require massive investments and often require the participation of the public sector in the form of shared investment or subsidies to be successful. One possibility of circumventing some of these barriers is to power existing dams that do not have hydroelectric generation capacity. This retrofit involves the installation of a turbine system that generates mechanical energy from flowing water and a generator system that converts mechanical energy to electrical energy. It also usually requires construction work on existing dam infrastructure. Similarly, retrofits to existing hydroelectric generation can significantly improve its generation capacity. In general, combining the powering of existing dams without generation with these existing generation retrofits can increase global generation capacity by between 6 % and 11 %, depending on the assumptions used for this estimation [30]. Although this approach to power generation and/or repowering could certainly be beneficial, global hydroelectric capacity would probably need to be expanded by at least 50 % and specifically dedicated to ammonia production in addition to other sustainable development goals. In addition, the use of hydroelectric power places geographical constraints on where ammonia can be produced, especially when locations where permits will be approved are overlaid with the appropriate topography.

As in the case of making ammonia from variable renewables, expanding ammonia production using hydroelectric power will require considerable increases in electrolysis manufacturing. Specific to hydropower, there does exist the possibility of pairing this generation with more efficient, less flexible solid oxide electrolysis (SOEC) technology. Although hydroelectric generation can exhibit seasonal or year-over-year production variation, its production on short timescales is more consistent relative to wind and solar resources. However, SOEC is lagging alkaline and PEM electrolysis in terms of deployment. To date, the largest operational SOEC systems have capacities less than 10 MW [160]. However, SOEC system production is rapidly scaling up, Bloom Energy has a GW-scale manufacturing facility in Fremont, California [39] and Topsoe is expected to begin manufacturing at a 500 MW-scale facility in Herning, Denmark [162]. In general, producing sustainable ammonia entirely from hydroelectric power and SOEC could require more than 1,000 GW each of generation and SOEC capacity. Even with the recent Bloom and Topsoe facilities, SOEC manufacturing needs to expand by orders of magnitude to materially contribute to sustainable ammonia manufacturing using hydropower.

2.4.3 Ammonia production from biomass

Sustainable ammonia production from biomass and other organic waste is the closest one-to-one replacement of fossil fuels from a process technology point of view. Once gathered, these resources are "dispatchable" in the same way as fossil fuels in the sense that their short-term availability does not depend on external factors. Making ammonia from biomass involves different upstream processes than using coal or natural gas to produce hydrogen, but downstream purification processes (e. g., water-gas shift, CO_2 removal) and ammonia synthesis are still able to operate in the same way. This is a distinct advantage for biomass-based ammonia production in comparison to the use of wind, solar, or hydroelectric power.

However, biomass-based ammonia production is highly dependent on the availability of these resources both in terms of quantity and spatial distribution. Currently, approximately 6.2 million terrajoules (TJ) of energy are produced from biomass and organic waste worldwide. This arises from both fuel and electricity generation. From a fuel perspective, approximately 2.3 million TJ of bioethanol and 1.2 million TJ of biodiesel are produced each year [376]. Figure 2.9 shows the geographic distribution of this fuel production. Meanwhile, biomass and organic waste are used to generate 2.7 million TJ of electricity [375]. In comparison, 200 million mt/y global ammonia production requires at least 7 million TJ assuming a theoretical best possible energy consumption of 35 GJ/mt

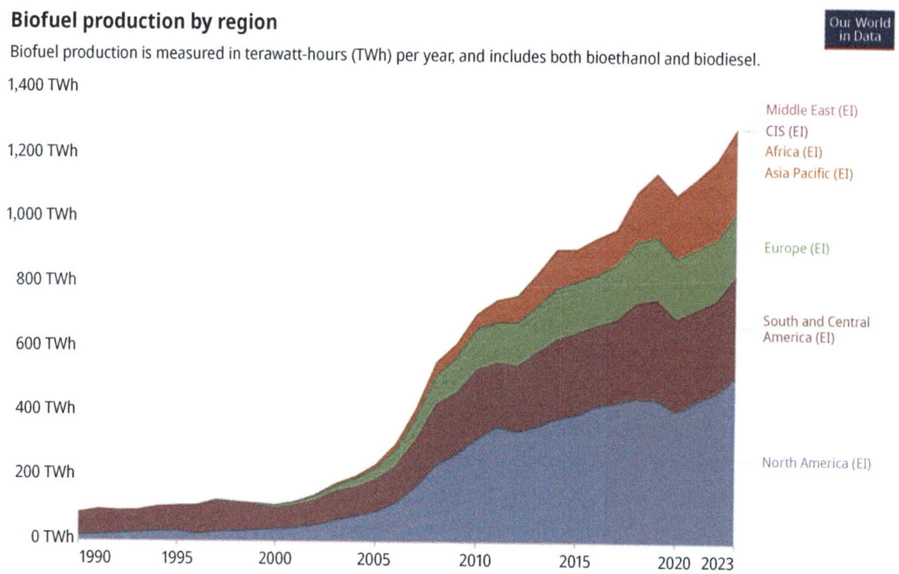

Figure 2.9: Biofuels production including ethanol and diesel by region. Fuel volumes are adjusted to TWh by energy content. The data in this figure is from Energy Institute – Statistical Review of World Energy (2024) with major processing by Our World in Data. Reprinted under Creative Common CC BY License.

in theoretical biomass-to-ammonia pathway. In other words, replacing all of the global current ammonia production with biomass-based production would not even be possible using all currently consumed bioenergy. From an ammonia production standpoint, global demand is expected to increase especially as ammonia use in the energy industry grows. This demand could reach 335 million mt/y by 2040 [5].

On the bioenergy availability side, some estimates indicate that bioenergy production could increase sustainably sixfold by 2050 [93]. However, this would require widespread global dietary changes to cooptimize human calorie intake while increasing land use for fuel crops. This is not likely to be a realistic scenario. In general, biomass-based production is not likely to be a dominant pathway toward making ammonia production more sustainable. This is compounded by the fact that there are other, more critical use cases for biomass and other organic waste. A considerable amount of biomass is dedicated to ethanol production, especially in the US and Brazil. This ethanol is currently predominantly blended with gasoline to reduce the carbon intensity of consumer vehicle transportation. In both countries, the ethanol industry is subsidized by the federal and state governments and has a significant economic impact on agriculture and farm-adjacent communities. Redirecting corn (in the US), sugarcane (in Brazil), and other bioresources to the production of ammonia would significantly disrupt multiple aspects of the current biofuel supply chains in those countries and others that import bioethanol.

The need for sustainable aviation fuel (SAF) further complicates the case for making ammonia from biomass. Aviation, especially of the long-haul variety, will require carbon-based fuels for the foreseeable future [336]. In contrast, ammonia of course does not contain any carbon, and its production does not necessitate carbon feedstocks. The projected use of bioenergy in near-term SAF production will already likely put some stress on the existing biomass and biofuel supply relative to its current demands. Furthermore, SAF can be specifically produced from bioethanol through alcohol-jet processes [396]. This would enable the existing ethanol industry to remain intact while accessing an even larger market for their products. SAF production fits much more synergistically with the existing bioenergy landscape than ammonia production, while simultaneously there are fewer viable alternatives for its production.

Another challenge in using biomass and other organic waste to make ammonia is the inherently distributed nature of these feedstocks. Biomass from dedicated energy crops or sourced as waste from food crops is spatially distributed over multiple acres of farmland and often multiple different farms. Furthermore, solid biomass contains up to 6.5 % hydrogen, which is a considerably lower hydrogen density than 25 % contained in methane [178]. This low hydrogen density limits the distance over which it is economical to transport biomass before upgrading its value density via chemical processing. This results in a network of smaller-scale production facilities. For example, the average bioethanol production facility in the United States has a capacity of 2.6 barrels of oil equivalent (boe) per day [138]. This is 95 % smaller than the average US oil refinery in terms of boe/d production capacity. On a hydrogen basis, the average US bioethanol plant could produce 50,000 mt/y hydrogen (or approximately 286,000 mt/year ammonia) based on its corn consumption. This somewhat scaled-down production further underscores the importance of making smaller-scale ammonia synthesis processes economically viable.

Finally, there are some broader sustainability concerns with the use of biomass and other organic waste as a feedstock for energy generation or chemical production. The magnitude of these concerns depends on the type of biomass or organic waste that is used. In general, from crop-based biomass, the required intensity of land, water, and fertilizer (which could include ammonia) must be taken into account rather than only the carbon intensity of the resulting ammonia. This is also complicated by the tension between dedicating crops for chemical feedstock versus using them for food, especially as the global population increases. Meanwhile, biomass or organic waste processing routes that involves methane as an intermediate, for example, fermentation of municipal solid waste, must be monitored carefully for methane emissions. Methane has 84 times the global warming potential of CO_2 over a 20-year period, and its emissions could offset the overall sustainability of the production of ammonia through this pathway rather than with fossil fuels.

3 Ammonia synthesis catalysis

This chapter discusses the conventional ammonia synthesis catalyst and the role of promoters in the context of the ammonia synthesis reaction. The chapter also discusses potential novel ammonia synthesis catalysts that allow the synthesis reaction to occur under milder operating conditions.

The reactor is considered the heart of a chemical process. Since most reactions are thermodynamically limited by equilibrium at the reaction temperature and pressure, it is required to devise a separation unit followed by the reactor in the synthesis loop. According to engineering heuristics, if the single-pass conversion is less than 95 %, a separation unit and a recycle line would be necessary [386]. For the case of ammonia production via the Haber–Bosch process (Figure 3.1a), the synthesis loop is operated at elevated pressure (100 bar and above) [222, 171]. The reaction temperature is typically controlled by adjusting the inlet synthesis gas temperature and space velocity. In the synthesis loop illustrated in Figure 3.1b, the synthesis gas is preheated to around 400 °C before entering the reactor. The gas temperature increases along the length of the reactor due to the exothermic reaction, with the outlet temperatures usually exceeding 520 °C. Such elevated pressures and temperatures are required to push the reaction toward the formation of more ammonia at a faster rate. Under such conditions, 15–20 % single-pass nitrogen conversion is expected, which requires the separation of unreacted hydrogen and nitrogen.

The equilibrium limitation significantly influences the design of the Haber–Bosch synthesis loop. A separation unit and a recycle line (with a recycle compressor) are required to improve the overall reactant conversion to over 99 %. To prepare the synthesis gas for condensation separation, the reactor exhaust is directed towards a chain of heat exchangers to cool the gas. Typically, for higher-pressure operation (greater than 200 bar), refrigeration is not required because the partial pressure of ammonia is high enough to condense ammonia at ambient temperature. More recent ammonia synthesis processes are optimized to operate at lower pressure (70–150 bar), so a refrigeration system is required to reduce the temperature to approximately −15 to −20 °C for efficient phase-change separation of ammonia. Once ammonia is separated, the synthesis gas (majorly hydrogen and nitrogen at a stoichiometric ratio) is returned to heat recovery and preheating before being introduced into the reactor using the recycle line.

Innovative reaction and separation approaches and optimization of the ammonia synthesis loop are essential toward maximizing the viability of sustainable ammonia production. At lower pressures and temperatures, the cost of capital decreases, which favors distributed manufacturing. However, reduced conversion at lower temperature and pressure requires greater recirculation compression work to sustain the reaction rates. To resolve this issue, advances in both the reaction and the separation parts of the Haber–Bosch process have been pursued as schematically illustrated in Figure 3.2. Catalysts that are active at lower temperatures have been

https://doi.org/10.1515/9783110762525-003

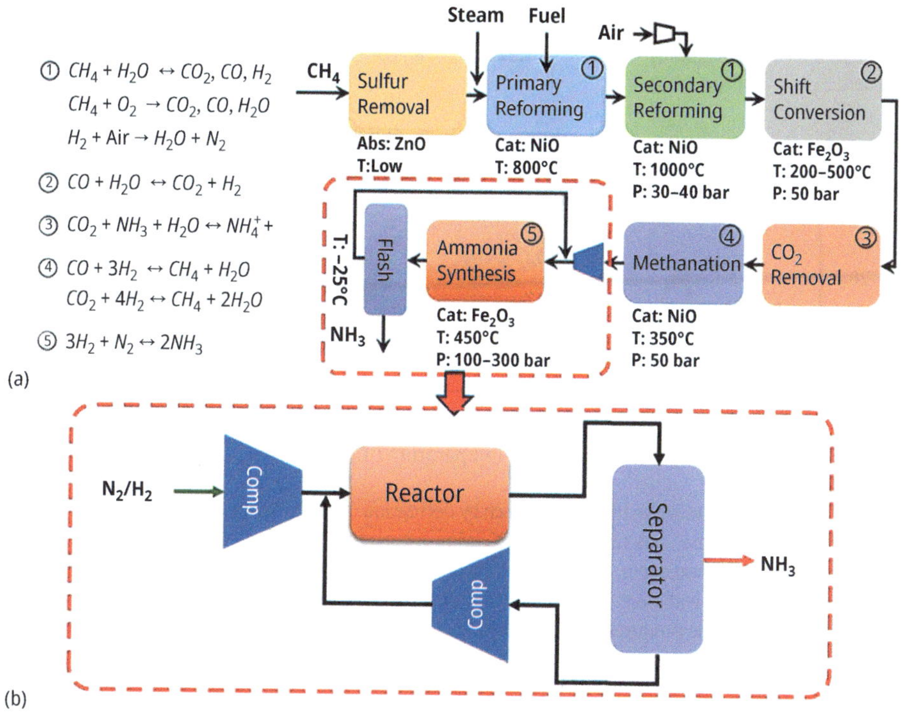

Figure 3.1: (a) Conventional ammonia production process from natural gas requires extensive pretreatment and processing to produce H_2 and N_2 at high purity. This pretreatment consists of sulfur removal, primary reforming, secondary reforming, shift conversion, CO_2 removal, and methanation. The gas then is compressed into the Haber–Bosch synthesis loop. (b) Key components of Haber–Bosch synthesis loop are a feed compressor, a reactor to convert H_2 and N_2, a recycle compressor to return unreacted gas back to the reactor. Reprinted with permission from Bosong Lin et al., Perspective on intensification of Haber–Bosch to enable ammonia production under milder conditions, ACS Sustainable Chemistry & Engineering, Volume 11, Issue 27, pp. 9880–9899. Copyright 2023 American Chemical Society.

pursued for more than a century [327]. In addition, the development of robust catalysts that can withstand fast heating cycles and tolerate humidity is of importance because ammonia manufacturing from variable renewable energy will be subject to the intermittency and impurity of the feedstock. Another interesting concept that can transition the current centralized Haber–Bosch to more sustainable routes is improving costs with advanced modularized multifunctional reactors that can improve efficiency while reducing costs. For separation purposes, selective nonthermal separations are of interest, as replacing phase change separation has the potential to increase the efficiency of a chemical process by an order of magnitude [342]. This chapter will mainly cover recent efforts on ammonia catalysis and advanced reactor designs. The next chapter will focus on possible improvements to the separation process.

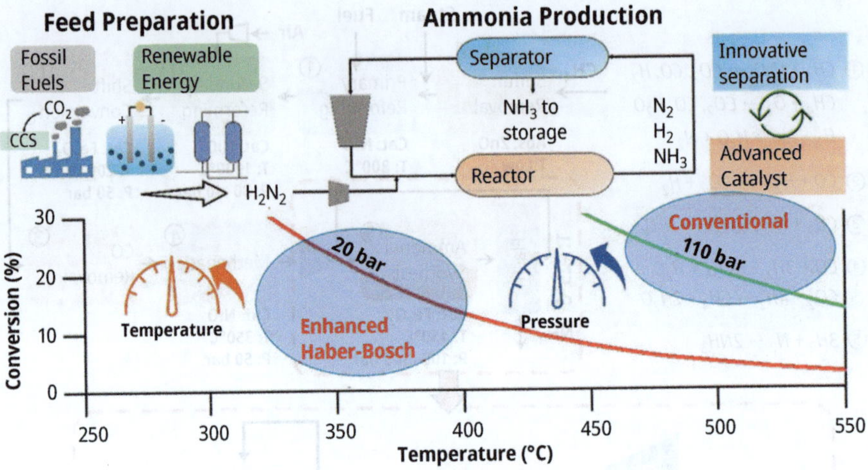

Figure 3.2: Pathways to improve the Haber–Bosch by decreasing pressure and temperature from at least 110 bar and 450 °C to 20 bar and 325 °CC while achieving comparable single-pass conversions include designing better catalysts and deploying more efficient separations of ammonia from unreacted hydrogen and nitrogen. Reprinted with permission from Bosong Lin et al., Perspective on intensification of Haber–Bosch to enable ammonia production under milder conditions, ACS Sustainable Chemistry & Engineering, Volume 11, Issue 27, pp. 9880–9899. Copyright 2023 American Chemical Society.

3.1 Catalyst selection

The tremendous amount of research conducted on the ammonia synthesis reaction has made this reaction known as the bellwether of heterogeneous catalysis [45]. The lessons learned from a century of research can guide the way toward the exploration of more active catalysts. Thus, it is imperative to understand the ammonia synthesis reaction and to learn what makes a good ammonia catalyst for nitrogen fixation and what limits us to make a more active catalyst.

3.1.1 Traditional catalyst and the role of promoters

In one of the early demonstrations of ammonia production, Haber and Bosch adopted a closed synthesis loop and used osmium as a catalyst [120]. Osmium was partially active and stable and required high pressure and temperature to achieve reasonable production rates. For large-scale processing, the abundance and cost of osmium were the major concerns. Hence, the fundamental characteristics of a suitable ammonia catalyst were set to be activity, stability, cost, and abundance. Motivated by Ostwald, Nernst, and Haber experiments on ammonia decomposition, Mittasch and coworkers tested iron catalysts for ammonia synthesis [258]. They noticed that iron performance improved markedly after the addition of impurities (i. e., promoters) to the catalyst formulation. The goal of such experiments was to facilitate the formation of iron nitride, which was considered to be the intermediate for the ammonia synthesis reaction; some impurities showed improved activities, while others were poisonous or led to catalyst deactivation. Eventually, the conventional iron-based catalyst was developed and refined through several decades of research and development, composed of ca. 94 % iron oxide and the balance being a combination of oxides (K_2O: 0.8 %, Al_2O_3: 2.3 %, CaO: 1.7 %, MgO: 0.5 %, SiO_2: 0.4 %) [49]. This mixture is fused into a single phase in a furnace and then ground into smaller pieces (3–5 mm). The introduction of some of the promoting components (e. g., Al, Ca, Si, and Mg) was found to result in improved textural properties, while the addition of other impurities (e. g., K) improved the kinetics of ammonia synthesis for optimal loading [197, 94]. A major observation after the addition of promoters was the fact that the catalyst was significantly more stable in the presence of impurities in the synthesis gas; a commercially promoted iron catalyst has been shown to demonstrate reasonable activity for more than 10 years.

3.1.2 Ammonia synthesis reaction

The ammonia synthesis reaction at 300 K and 1 bar is exergonic; thus, from a thermodynamic point of view, the reaction at this condition is feasible [384]. However, theoretical and experimental studies on surface adsorption energies indicate that at 300 K, reaction intermediates (NH_x, $x = 0, 1$, and 2 in the case of ammonia) bind too strongly to the catalyst active sites. This necessitates operating at higher temperature (>650 K) to reduce adsorption energies (E_N). However, at elevated temperatures, the reaction has positive free energy. Substantial operating pressure above 100 bar is therefore required to make the reaction exergonic again. Thereby, low-temperature/low-pressure ammonia synthesis is feasible only for elements which show strong activation of nitrogen and bind amines more weakly in the reaction condition. Assuming that nitrogen dissociation is the rate-limiting step in transition metal-mediated ammonia catalysis, all conventional 3d and 4d transition metal catalysts utilized to date show a linear energy scaling relation between the transition state energy required for nitrogen dissociation (E_{N-N}) and the adsorption energy of nitrogen at the catalyst sites (E_N), as illustrated in Figure 3.3b. E_{N-N} is a representative of the activation energy of the catalyst determining the rate of dissociation, while E_N represents the surface coverage by adsorbed nitrogen. When the rates are plotted versus E_N, 3d- transition metals demonstrate a volcano-type relationship, with Sabatier optimum catalysts at the top of the volcano (Figure 3.3c).

According to the Sabatier principle in heterogeneous catalysis, a “good” catalyst is one in which the interactions between reaction sites with reactants and products are “just right”; neither too weak (capable of activating the reactants) nor too strong (not tightly adhering to the products). This theory is used as a predictive tool to design better catalysts [251]. As illustrated in Figure 3.3a, the catalytic activities of the stepped (211) transition metal surfaces are limited by too strong bonds for reactants or products. The

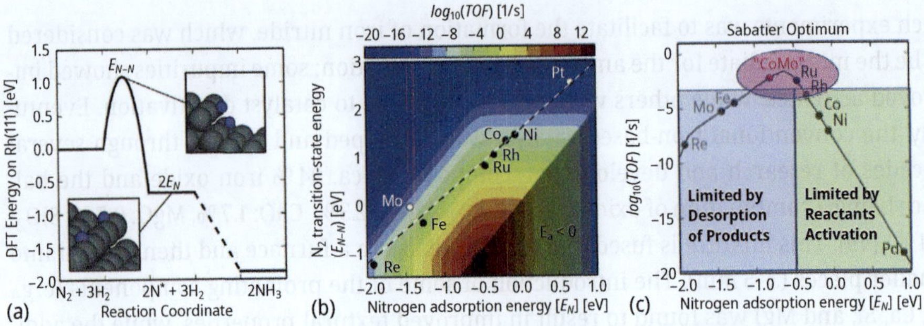

Figure 3.3: (a) Synthesis rate calculated using a mean-field kinetic model as a function of transition-state energy and adsorption energy of nitrogen on face-centered cubic Rhodium (FCC Rh). (b) Ammonia synthesis rates as a function of nitrogen transition state-energy and nitrogen adsorption energy for various transition metals display a scaling line. (c) Ammonia synthesis rates vs. nitrogen adsorption energy display a volcano-type relationship for 3d- transition metals. Reprinted from Journal of Catalysis, Volume 328, Andrew J. Medford et al., From the Sabatier principle to a predictive theory of transition-metal heterogeneous catalysis, pp. 36–42, Copyright 2015, with permission from Elsevier.

contour map in 3.3b clearly illustrates the effect of the scaling relation on the reaction rates. Among the most studied transition metals, Fe and Ru with more balanced absorption energies have shown the highest turnover frequencies, assuming that they experience similar influence because of electronic and geometric promoters. These findings are taken into account for the rational design of catalysts with alloyed transition metals from different sides of the volcano such as Co and Mo showing improved performance, which was attributed to the presence of two active sites with balanced adsorption energies for reactants and products [168]. Nonetheless, the catalyst performance improvement still remained limited to those of typical transition metals.

3.1.3 Conventional ammonia catalyst

Since the emergence of the first catalysts for synthetic nitrogen fixation by Haber and later by Mittasch, improving the catalyst has been a holy grail in the study of heterogeneous catalysis. One of the promising routes identified was to use more active catalyst materials, and most prominently ruthenium (Ru) became the focus of the second generation of ammonia synthesis catalyst under milder conditions. In the comprehensive efforts by Mittasch, Ru was not actually found to be sufficiently active. Unlike iron, they found that Ru did not show any improved activity when promoted; interestingly it was reported that Ru instead (unlike others tested in this work) showed improvement with the support. The first detailed study of Ru-based catalysts was by Aika et al., which conducted kinetics and isotope studies and suggested that nitrogen chemisorption at the catalyst sites was the rate-limiting step [258], but there was not much attention paid towards the odd effect of hydrogen partial pressure on the synthesis rates [15]. The academic research on Ru-based catalyst was very limited (except for some of the pioneering work by Ozaki from Tokyo Institute of Technology) until MW Kellogg took the first step in 1992 to produce 600 tonnes/day ammonia plant using carbon-supported Ru catalyst [265]. Since then, the research on Ru-based catalysts has been majorly focused on improving the support-effect, promotion, or developing novel and simpler methods for improving the catalytic site dispersion over the support. Carbonaceous support has been most widely used. The most effective promoters studied to date are alkali and alkaline earth metals and their oxides [315, 14, 369, 165]. There are two major objectives in promoting of Ru. First, it is widely known that saturation of the catalyst sites with hydrogen is found to be a challenge with Ru that can be partially resolved with promotion. Second, Ru cannot form a stable nitride species; thus a strong electron promoter is needed to facilitate breaking the triple nitrogen bond.

A major challenge with Ru though is the cost and the risk of its supply chain. Ru belongs to the platinum group metals and is relatively rare on earth. What makes it even more challenging to develop Ru-based catalysts is the choice of the precursor [117], as the Ru precursor can partially alter the surface properties of the catalyst [429]. The same catalyst derived from different precursors has displayed different production rates. There have been more recent efforts to use innovative catalytic materials based on Ru, which are still at the early stages of research and will be discussed in the next section. Figure 3.4 summarizes the recent efforts in improving Ru-based ammonia synthesis. The data is categorized based on two key variables: synthesis temperature and Ru loading. Most catalyst activities are reported under 30 bar, while it is critical to evaluate hydrogen

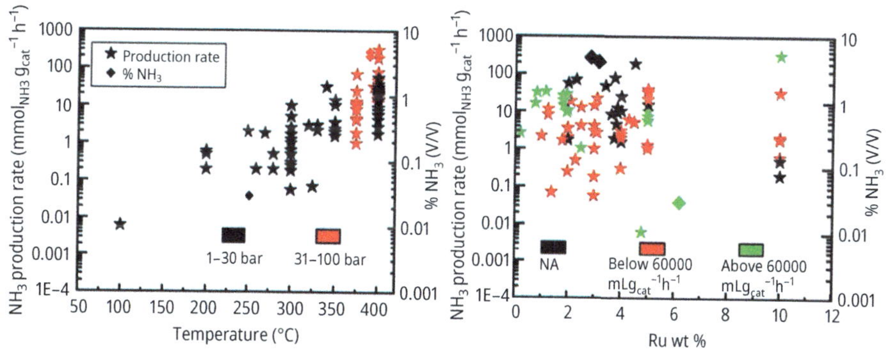

Figure 3.4: Ammonia synthesis activity over different Ru-based catalysts. (a) production rate and outlet concentration are plotted against the lowest temperature reported for the respective catalysts. The colormap legend displays the pressure range (bar). (b) The figure is plotted for the same results where the production rate is shown against Ru wt.%, with the goal to highlight the trend to reduce the loading of Ru on inert porous supports for reduced costs. The color map shows the space velocity (mL $g_{cat}^{-}1$ $h^{-}1$). Table 3.1 summarizes more details of the catalysts displayed above. Reprinted with permission from Bosong Lin et al., Perspective on intensification of Haber–Bosch to enable ammonia production under milder conditions, ACS Sustainable Chemistry & Engineering, Volume 11, Issue 27, pp. 9880–9899. Copyright 2023 American Chemical Society.

Table 3.1: Summary of catalysts, Ru loading, and activities of the catalyst reported under various operating conditions. Reprinted with permission from Bosong Lin et al., Perspective on intensification of Haber–Bosch to enable ammonia production under milder conditions, ACS Sustainable Chemistry & Engineering, Volume 11, Issue 27, pp. 9880–9899. Copyright 2023 American Chemical Society.

Catalyst	Lowest reported temperature [°C]	%NH_3 [V/V]	Production rate [$mmol_{NH_3}$ g_{cat} h^{-1}]	Pressure [bar]	Space velocity [mL g_{cat}^{-1} h^{-1}]	Ru content [wt.%]	Ref.
Ba-Ru-Li/AC	100		0.0062	1	62,400	4.8	[426]
Ru/BaO-CaH_2	200		0.5	10	36,000	10	[129]
Ru/Ba-$Ca(NH_2)_2$	200		0.6	10	36,000	10	[129]
Ru/$Ca(NH_2)_2$	200		0.2	1	36,000	10	[129]
CsRuBaCCHT-PR	252	0.03		10	66,666	6.2	[267]
2.5Cs-10Ru/MPC-18	260		0.02	9	NA	10	[278]
Cs-Ru/MPC	260		0.02	9	NA	10	[278]
0.5Ba-10Ru/MPC	280		0.5	9	NA	10	[279]
Ru/$BaCeO_3$-xNyHz	300		5.0	9	36,000	4.5	[192]
Ru/$La_{0.5}Ce_{0.5}O_{1.75}$	300		10.0	10	72,000	5	[283]
Ru/Pr_2O_3	300		1.25	10	18,000	5	[150]
Ru/CeO_2	300		1	10	18,000	5	[150]
BaRu-graphene	325		0.07	10	36,000	1.4	[425]
Ru/Ti-Ce-S	330		3	10	36,000	3	[407]
Ru/$Ca(NH_2)_2$	340		32	9	36,000	10	[153]
Ru/BZY10-1200	350		2	10	36,000	2	[341]
K_Ru/BZY10-1200	350		3.5	10	36,000	2	[341]
Cs_Ru/BZY10-1201	350		3.6	10	36,000	2	[341]
Ru-Ba-Cs/Sibunit	350		1.54	7	NA	4	[163]
Ba-K/Ru-MC	375		70	100	NA	2.3	[433]
Ru/$La_2Ce_2O_7$	375		25	100	NA	4	[123]
2 %Ba-2.5 %Ru/CeO_2-MS	375		4.2	38	24,000	2.5	[230]
2.5 %Ru/CeO_2-CS	375		1	30	70,000	2.5	[231]
Ru/$BaTiO_3$	375		8.29	50	NA	3.63	[398]
Ru/CeO_2	375		4.55	50	NA	3.82	[398]
Ru/MgO	375		1.88	50	NA	3.74	[398]

Table 3.1 (continued)

Catalyst	Lowest reported temperature [°C]	%NH_3 [V/V]	Production rate [$mmol_{NH_3}$ g_{cat} h^{-1}]	Pressure [bar]	Space velocity [mL g_{cat}^{-1} h^{-1}]	Ru content [wt.%]	Ref.
1.25 % Ru/$BaCeO_3$-a	375		8	30	24,000	1.25	[215]
4 %Cs-1.25 % Ru/$BaCeO_3$-a	375		11.5	30	24,000	1.25	[215]
Ru/$BaZrO_3$	375		10.18	30	NA	4	[399]
2.5 %Ru/CeO_2-TPAOH	375		13	30	2,4000	2.5	[214]
SmRu/CCA_3	390	4.8	–	100	NA	2.83	[221]
SmRu/Al_2O_3	390	4.1	–	100	NA	3.12	[221]
Ba-Ru/BN	400		187.5	100	NA	4.5	[167]
Ba/Ru-N-MC	400		79	100	NA	3.75	[239]
Ba–Ru/CNTs-D	400		48	100	NA	3.4	[239]
Ba-Ru/AC-G	400		312	100	70,000	10	[218]
Ru/$BaTiO_{2.7}H_{0.3}$	400	30.36	50	66,000	0.81		[358]
Ru/$SrTiO_{2.6}H_{0.4}$	400	15.74	50	66,000	0.75		[358]
K-Ru/Ba-ZrO_2-1.0	400		22.82	50	NA	4	[224]
Ru/Y-LDO	400		12.35	30	NA	3.85	[276]
Ru/MgAl-LDO	400		9.1	30	NA	3.86	[276]
Ba-Ru/gc-Al_2O_3	400	5.6	10	60,000	5		[220]
Ru-Ba/Al_2O_3-980	400		7.217	10	60,000	5	[219]
Ru-Cs/MgO-MIL	400		22.29	10	24,000	3.1	[213]
Ru/C_inCs	250		2.0	10	48,000	10	[335]
Cs-Ru/SGCNT	270		1.8	1	36,000	10	[280]
Ba-(Ru/AC)o	400		38	10	36,000	5	[277]
Ru/3TiCN/ZrH_2	350		10.3	10	60,000	2	[432]
Ru/0.2Cs-MgO	300		1.0	10	24,000	3	[229]
Ce/Ba/Ru NCs	350		13.92	10	60,000	1.83	[303]
Ru/$Ti_{0.18}$-Ce	400		18.912	10	36,000	2.1	[211]
2Cs-Ru/r-CeO_2	320		2.9	10	18,000	4	[242]
Ru/CaBTO	300		0.5	1	36,000	2.3	[143]

poisoning at higher pressure which is lacking in the majority of the literature. The average Ru loading was between 3–5 %, and only two relatively active catalysts with lower Ru loading (<1 wt.%) are reported [358]. To date, the lowest temperature of ammonia synthesis reported is 100 °C at 1 bar.

3.2 Catalysts for synthesis under milder conditions

Innovative approaches to develop better catalysts have focused on improving the support-active site promoting effect, enhancing catalyst activity with multisite catalysis [307], and reducing catalytic sites by restricting catalyst precursors in a confined support. The catalysts discussed in the following are still in the early stages of research and development (technology readiness level 1–5).

3.2.1 Nitrides

Initial efforts to synthesize better ammonia catalysts were actually initiated by alloying transition metals and synthesizing ternary nitrides. Jacobsen synthesized catalysts of ternal metal-molybdenum (Mo) nitrides where metal can represent iron (Fe), Co, and nickel (Ni) for ammonia synthesis and found that cesium-promoted Co_3Mo_3N was more active than the multipromoted iron catalyst [166]. The mechanism of ammonia synthesis using binary and ternary nitrides is considered to follow a Mars–van Krevelen mechanism, as shown in Figure 3.5 [68]. Based on this mechanism, lattice nitrogen plays a role as an active species in nitrogen hydrogenation, rather than being an inert species in the catalyst, as illustrated in Figure 3.5. This was further investigated and validated by Daisley and Hargreaves using isotopic exchange experiments [146]. Recently, it was shown that dissociated hydrogen could directly react with the nitrogen in the lattice of nickel-loaded lanthanum nitride (LaN) and produce ammonia and form nitrogen vacancy sites under mild conditions (400 °C at 1 and 9 bar) [411].

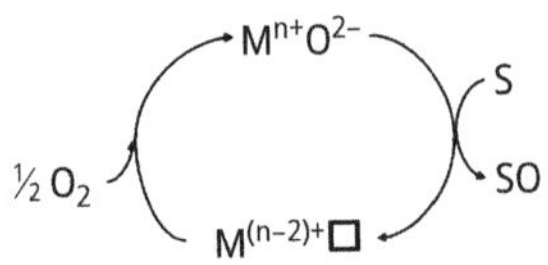

Figure 3.5: Nitrogen based chemical looping for ammonia synthesis. The proposed mechanism suggests that ammonia synthesis is catalyzed by metal nitride via Mars–van Krevelen mechanism. Reprinted from [68] under Creative Common CC BY License.

Wang et al. also proposed a Co-Ni-C catalyst for enhanced ammonia synthesis [397]. In this case, Co single-atom sites supported on nitrogen-doped hollow carbon spheres were studied and enhanced ammonia synthesis rates were observed, although Co is known to be not an active catalyst for ammonia synthesis [397]. The proposed catalyst was synthesized by wet impregnation of cobalt phthalocynine (CoPc) on nitrogen-doped hollow carbon spheres. First, nitrogen-doped carbon hollow spheres were synthesized by reacting ammonia, ethanol, and water, followed by the addition of tetraethyl orthosilicate, F127 ($PEO_{106}PPO_{70}PEO_{106}$), resorcinol, and formaldehyde, melamine. After sequential addition and soaking, the product was filtered, washed, and heated up to 700 °C under Ar flow to carbonize the spheres. Later, CoPc dissolved in dimethylformamide was impregnated in nitrogen-doped carbon hollow spheres to produce single atom Co catalysts (Co-N-C), as shown in Figure 3.6a-d. Authors claimed that Co was found to be present in single atom clusters rather than in larger clusters of nanoparticles; hence, Co catalyst sites were not visible in the TEM results. However, high-angle annular dark-field scanning TEM (HAADF-TEM) displayed the atomic dispersion of Co over the carbon support, as shown in Figure 3.6e. X-ray absorption also did not detect any Co-Co coordination peak, suggesting that Co is present in its single-atom structure (Figure 3.6f and 3.6g). Through controlled reaction and temperature-programmed tests, the authors actually confirmed that pyridine nitrogen can be hydrogenated to produce ammonia via chemical looping.

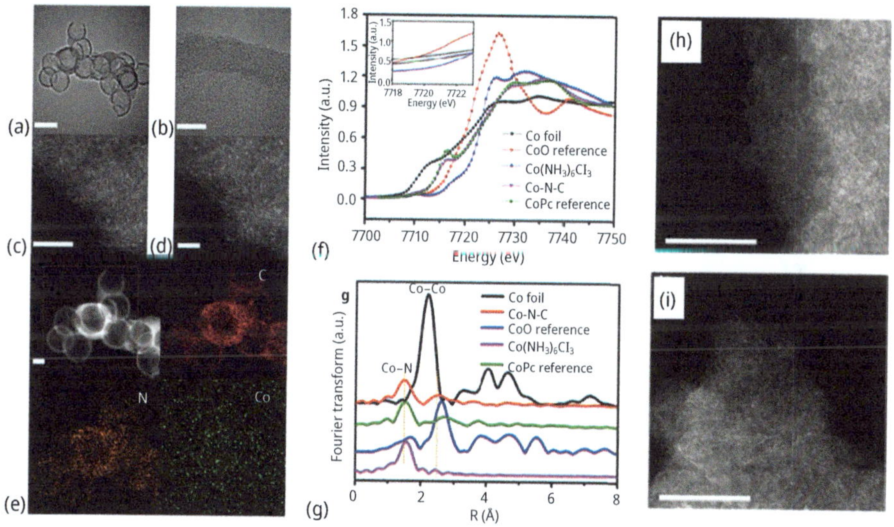

Figure 3.6: (a)–(d) High-resolution TEM images of Co-N-C displays a uniform distribution of Co single atoms (not visible) on uniform nitrogen doped carbon (130 nm average size). Scale bars are 200 nm, 10 nm, 5 nm, and 2 nm for a, b, c, and d, respectively. (e) HAADF-STEM images of Co-N-C confirm the uniform distribution and existence of C, N, and Co atoms in the spheres. (scale bar 50 nm). (f) and (g) Normalized Co K-edge XANES and Co K-edge EXAFS spectra of Co-N-C that were used to estimate the valence states of Co species. (h) and (i) The HAADF-TEM images of spent Co-N-C catalyst display a uniform distribution of single-atom Co and no sign of sintering. Reprinted from [397] under Creative Common CC BY License.

According to the authors, the mechanism of synthesis is through a chemical looping process, as illustrated in Figure 3.7. It is proven through a variety of density functional theory (DFT) and experimental work that Co is not active enough to break the triple bonds of nitrogen. Over the Co-N-C catalyst, nitrogen is first adsorbed on Co single atom sites, and forms Co_1-$N_{3.5}$, confirmed by argon temperature-programmed desorption tests. The Co_1-$N_{3.5}$ site is then activated by adsorption of hydrogen to form an intermediate (*NHNH), indicating that the proposed mechanism does not fall under the nitrogen dissociation route. The intermediate is then further hydrogenated to form *NH-NH_3 and subsequently to *NH-NH_4 on Co_1-$N_{3.5}$ sites, which later evolves back to *NHNH and releases two molecules of ammonia. This mechanism is primarily feasible due to the absence of B_5 sites in the single-atom Co_1 $N_{3.5}$ sites. It also justifies nitrogen adsorption, hydrogenation, and breaking the *NH_2-NH_3 bond is more favorable to the energy required to break nitrogen bonds and following the dissociative nitrogen mechanism. As shown in Figures 3.7b and 3.7c, the increased synthesis rates are due to the weak pyridine N attached to the Co site, which can easily undergo hydrogenation and form ammonia and leave a N vacancy. This vacancy can be later replenished by nitrogen in the synthesis gas at the reaction temperature and pressure. Follow up work also confirmed the unique contribution of weak N pyridine in the atomically dispersed catalyst to synthesize ammonia using Co [431, 262].

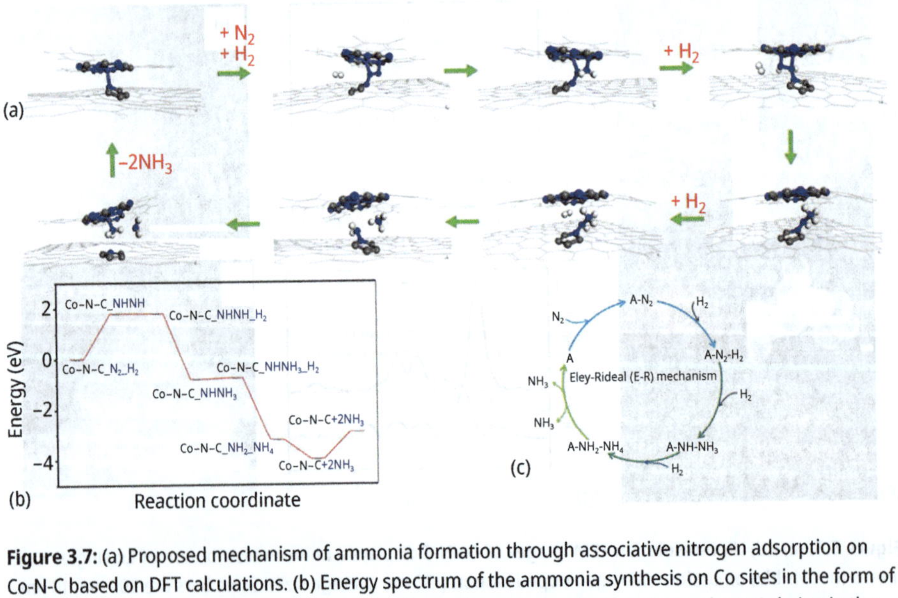

Figure 3.7: (a) Proposed mechanism of ammonia formation through associative nitrogen adsorption on Co-N-C based on DFT calculations. (b) Energy spectrum of the ammonia synthesis on Co sites in the form of Co_1-$N_{3.5}$. (c) Ammonia synthesis via dynamic cyclic site through chemical looping pathway (x being in the range of 0–1.5 and V_N^* stands for a nitrogen vacancy). Reused from [397] under Creative Common CC BY License.

3.2.2 Electride

Facilitating electron donation in transition metals has been considered a promising path for enhanced ammonia synthesis under milder conditions, and inorganic electrides are prominent examples of this approach. Kitano et al. first reported Ru-loaded electride of $[Ca_{24}Al_{28}O_{64}]^{4+}(e^-)_4$ (Ru/C12A7:e$^-$) with high electron donation properties, which led to an increase in the synthesis rates by an order of magnitude, compared to previously reported Ru-based catalysts [190]. A unique feature of the proposed electride is that it is abundantly available and stable at elevated temperatures. A schematic of Ru/C12A7:e$^-$ is illustrated in Figure 3.8a. C12A7:e$^-$ has an electron trapped in the O^{2-} vacancy of the CaO crystal. These nanostructured cages with high electron density can exchange electrons with hydride ions from hydrogen in the synthesis gas at atmospheric pressure. To close the cycle, the hydride ions can desorb as hydrogen at 400 °C and C12A7 can return to its positively charged framework through an irreversible reaction. This reversible hydrogen uptake and release is advantageous for the Ru-loaded catalyst because it can effectively prevent hydrogen poisoning of the Ru catalyst sites, which is a major obstacle with Ru-based catalysts for ammonia synthesis; the excess hydrogen from Ru spills over the electride and is entrapped as hydride ion. Activity tests of Ru/C12A7:e$^-$ vs. other conventional Ru-based catalysts (shown in Figure 3.8-c) confirm that strong electron donation of electride facilitates synthesis while reducing the hydrogen poisoning of Ru sites [128].

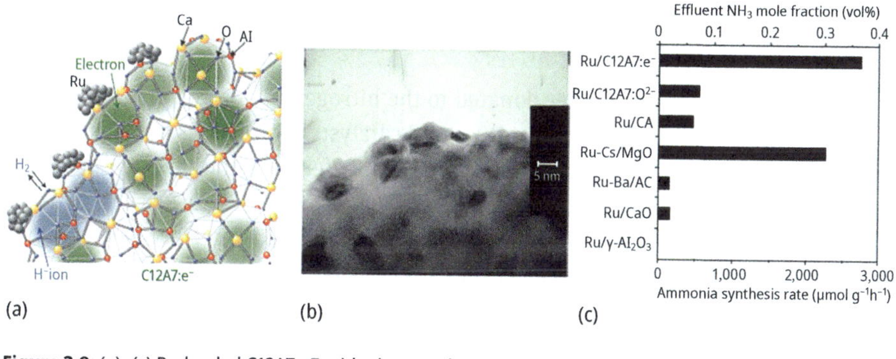

Figure 3.8: (a)–(c) Ru-loaded C12A7:e$^-$ with electron-dense structure distributed uniformly over sub-nanostructured cages that can exchange electrons and hydrides reversibly to promote ammonia synthesis. (b) TEM showing the Ru-loaded C12A7:e$^-$ with 0.3 % Ru loading. (c) Ammonia synthesis rates of Ru-loaded C12A7:e$^-$ vs. other conventional Ru-based catalysts at 1 bar and 400 °C at 60 mL min^{-1} synthesis gas (H_2/N_2=3) over 200 mg catalyst. Adapted from Kitano et al., Ammonia synthesis using a stable electride as an electron donor and reversible hydrogen store, Nature Chemistry, Volume 4, pp. 934–940, 2012 with permission from Springer Nature.

The proposed ammonia synthesis on Ru/C12A7:e$^-$ was investigated using spectroscopic and kinetic tests, as illustrated in Figure 3.9. Electrons are donated from the elec-

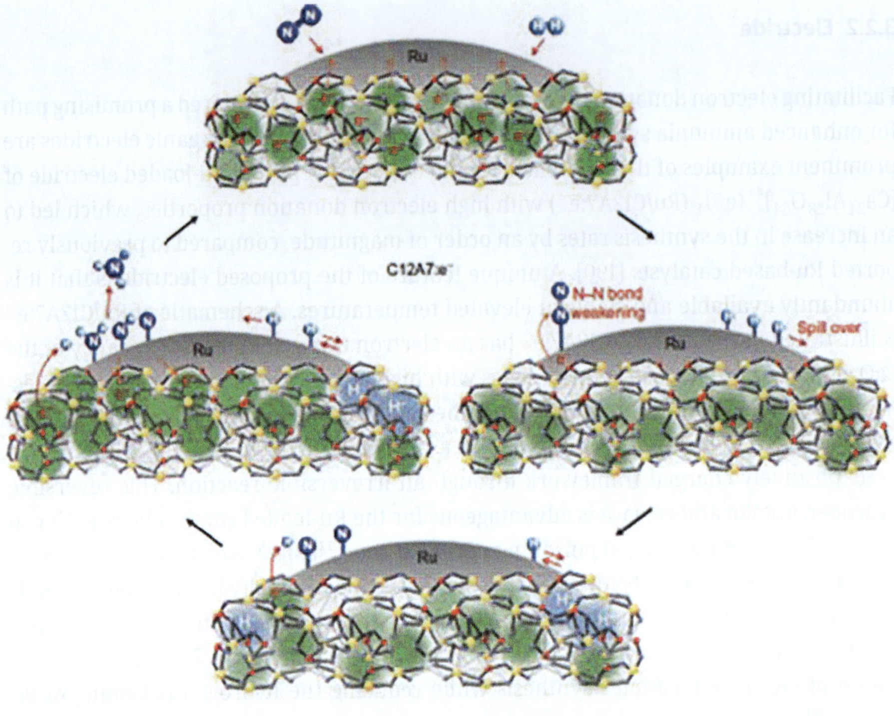

Figure 3.9: Proposed mechanism of ammonia synthesis over Ru/C12A7:e$^-$. Reprinted from Kitano et al., Ammonia synthesis using a stable electride as an electron donor and reversible hydrogen store, Nature Chemistry, Volume 4, pp. 934–940, 2012 with permission from Springer Nature.

tride to the Ru sites, which is later donated to the nitrogen π^*-antibonding orbital adsorbed on the Ru interface. Like other Ru-based catalysts, hydrogen is adsorbed on Ru sites quickly, but the spill over hydrogen is facilely transferred to the electride sites with electron deficiency, which prevents poisoning and overcrowding of the Ru sites. Hydride atoms at electride nanocages later react with nitrogen at the Ru interface (with weaker bonds) and will be replaced by an electron. This cycle continues, and the vacancy sites are being reversibly replenished with hydride and electrons under the reaction conditions. Hence, through this mechanism, the well-known reaction rate-limiting step for nitrogen hydrogenation shifts from nitrogen dissociation to ammonia synthesis using Ru supported on electrides [191].

Since the first demonstration of electrides, there have been numerous reports on the use of electrides of different chemistries. Notably, Kishida proposed Ru-loaded calcium cyanamide ($CaCN_2$) as an air-stable electride for ammonia synthesis [188]. $CaCN_2$ formed a quasi-electride structure that enhanced the electron donation of $CaCN_2$. Lu et al. designed air and water durable electride (Y_5Si_3) for ammonia synthesis [233]. Hayashi et al. developed a rapid thermal annealing method for treatment of C12A7:e$^-$ to obtain a conductive surface for improved electron donation [131]. Li et al. reported acid-durable lanthanide-based electrides with layered Ru (i. e., LaRuSi, LaRuSi, PrRuSi, and NdRuSi), in which Ru sites were exposed via selective chemical etching [212]. These catalysts are free of B5 sites, and adsorption energies of nitrogen and N are −0.7 and −0.8 eV, which are much lower than conventional Ru catalysts that relied on B5 sites to activate nitrogen. Recently, Tsubame BHB is paving the way to transition to commercialization of electride-based catalysts based on the intellectual property developed through electride research [140, 139, 141, 175]. Table 3.2 summarizes some of the reported activities of electride catalysts reported by Hosono et al.

Table 3.2: Summary of catalysts, Ru loading, and activities of the catalyst reported under various operating conditions. Reprinted with permission from Bosong Lin et al., Perspective on intensification of Haber–Bosch to enable ammonia production under milder conditions, ACS Sustainable Chemistry & Engineering, Volume 11, Issue 27, pp. 9880–9899. Copyright 2023 American Chemical Society.

Catalyst	Ru loading [wt. %]	Activity at different temperature [mol_{NH_3} g_{cat} h^{-1}] 260 °C	300 °C	340 °C	400 °C	Space velocity [mL g_{cat}^{-1} h^{-1}]	Activation energy [kJ mol^{-1}]	Ref.
Ru/C12A7:e^{-1}	1.8	193	745	2,021	2,760[a]	36,000	51 (320–400)	[190]
Ru/C12A7:e^{-1}	1.8	193	745	2,021	2,760[a]	36,000	91 (200–320)	[190]
HT-Ru/C12A7:e^{-1}	2.0			2,290		18,000	53 (320–400)	[152]
Ru/Ca_2N:e^{-1}	1.8	500	1,674	3,386		36,000	60 (250–340)	[189]
Ru/CeH_{2+x}	2.0	666	1,808	3,164		36,000	55 (260–340)	[259]
Ru/LaH_{2+x}	2.0	500	1,340	3,386		36,000	54 (260–340)	[259]
Ru/Y_5Si_3	7.8			818[b]	1,900	36,000	52 (260–340)	[233]
Ru/LaScSi	1.8				2,800	36,000	52 (260–340)	[403]
Ru/LaScSi	8.3				3,400	36,000	49 (260–360)	[403]
LaRuSi before EDTA				840	1,810	36,000	40.4 (320–400)	[212]
LaRuSi after EDTA				3,020	5,340	36,000	48 (320–400)	[212]
CaRuSi					60	36,000		[404]
Ru-Cs/MgO	6				3,400	36,000	73 (320–400)	[128]

3.2.3 Hydrides/Imides/Amides

Hydrides, imides, and amides were first studied for hydrogen storage applications. Temperature-programmed desorption of ball milled lithium amide ($LiNH_2$) and lithium amide-hydride combinations ($LiNH_2$-LiH and $LiNH_2$-2LiH) displayed an anomaly for $LiNH_2$ [56]. Among these samples, the decomposition of $LiNH_2$ was found to be initiated at 200 °C. At 374 °C ($LiNH_2$ melting point) and 421 °C, ammonia peaks were observed for $LiNH_2$. Prominent hydrogen and nitrogen peaks were observed at higher temperatures

(i. e., 391 °C and 441 °C) as a result of ammonia decomposition, whereas ammonia decomposition at such temperatures requires a very active catalyst. Later, this idea was pursued to design an ammonia decomposition catalyst based on metal hydride [70]. In particular, Guo et al. observed that ammonia decomposition rates were intensified in the presence of transition metals, which demonstrated the synergistic effect of Li_2NH on 3d-transition metals for ammonia decomposition [119]. Assuming that a good ammonia decomposition catalyst can be a good ammonia synthesis catalyst, the synergistic effect of imide/amide chemistry on transition metals was leveraged to design highly active composite catalysts of transition metals and LiH for ammonia synthesis [116, 118, 393]. At the reaction conditions and in the presence of synthesis gas (i. e., hydrogen), amide/imide transform to their hydride structure, so it is the hydride of the alkali and alkaline earth metals that can synergize transition metals for enhanced ammonia synthesis. Wang et al. reported for the first time that compositing 3d- transition metals with LiH show unprecedented catalytic behavior for ammonia synthesis under mild conditions [393].

The proposed mechanism of metal hydride-assisted ammonia synthesis is illustrated in Figure 3.10a [393]. According to spectroscopic measurements and X-ray diffraction studies, the presence of amide/imide structure in the composite catalyst was confirmed for the transition metal-LiH composite catalyst (i. e., Mn-LiH). Because LiH cannot

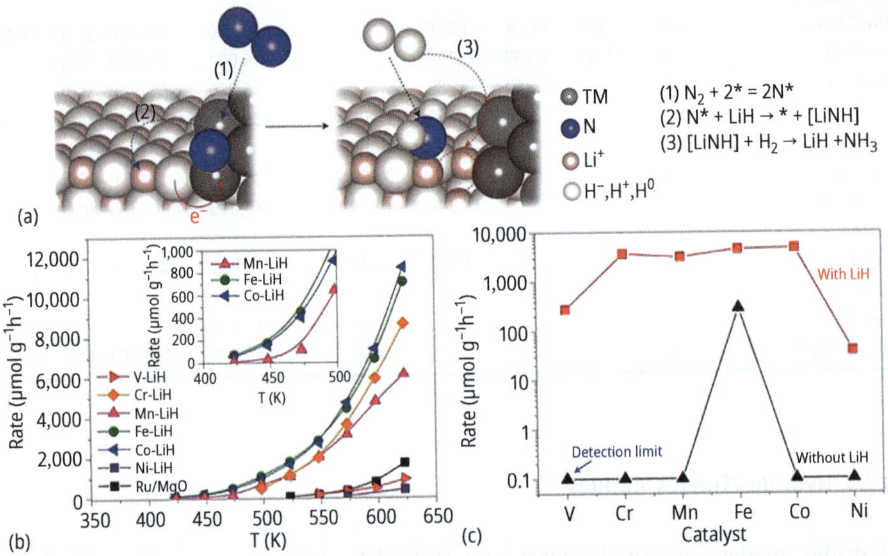

Figure 3.10: (a) Proposed mechanism of LiH-mediated ammonia synthesis on composite transition metal halides-lithium hydride structure. The hydrogenation of nitrogen species follows the three reactions provided. (b) and (c) comparison of the ammonia synthesis rates over TM-LiH catalysts at $H_2/N_2 = 3$ with a weight hourly space velocity of 60,000 mL g^{-1} h^{-1} and 10 bar. Adapted from Wang et al., Breaking scaling relations to achieve low-temperature ammonia synthesis through LiH-mediated nitrogen transfer and hydrogenation, Nature Chemistry, Volume 9, pp. 64–70, 2017 with permission from Springer Nature.

react with nitrogen under reaction conditions to form amide/imide structures, it was hypothesized that nitrogen species were supplied through the dissociation of the nitrogen molecule over MnNy (Mn exists in its nitride form at the reaction conditions). Hence, nitrogen is activated over the transition metal sites via reaction (1). Due to the strong reducing characteristic of LiH, nitrogen species are transferred from the transition metal (or their nitrides) to hydrogen-rich LiH to form Li-N-H compounds by reaction (2). At the reaction conditions, the heterolytic splitting of hydrogen and reaction with Li-N-H result in the production of ammonia and regeneration of LiH. Various 3d transition metals (i. e., V, Cr, Mn, Fe, Co, and Ni) were ball milled with LiH to synthesize composite catalysts and evaluated for activity tests in the temperature range of 150–350 °C and 10 bar. Almost all the catalysts showed promising results for ammonia synthesis (Figure 3.10b and c), while conventional catalysts did not show any activity, except iron, which showed an order of magnitude lower production rate, compared to Fe-LiH catalyst.

3.2.4 Metal Organic Frameworks

In a series of publications [238, 300] and a patent application [256], Luz et al. demonstrated the synthesis of Metal Organic Framework (MOF)-derived nanostructured Ruthenium catalysts. They showed that Ru-containing MOFs can be used to encage Ru within the defects of the HKAUST-1 MOF. Leveraging this technique allows the Ru catalyst sites to be tuned and adjusted by changing the organic struts in diverse structures of MOFs, which is a great tool that can be utilized to control the dispersion and catalyst particle size as well as optimize the synthesis rates. This tunability gives MOFs a unique advantage that can make them superior to conventional supports used in ammonia synthesis catalysts. The excellent dispersion of Ru in the defect-engineered MOF structure, followed by controlled pyrolysis of the catalyst, led to the synthesis of a highly dispersed Ru-based catalyst with enhanced synthesis rates at temperatures under mild conditions.

Luz et al. synthesized defected MOFs with ruthenium centers following the approach of Kozachuk et al. [195]. These defected MOFs of (Ru)HKUST-1 first contained 100 % benzene tricarboxylic acid (Hbtc). After thorough washing and purification, the RuMOF samples were evacuated under vacuum. The nanostructured catalysts were later developed in controlled pyrolysis under an inert environment (e. g., nitrogen) with a controlled temperature ramp at 900 °C, resulting in the formation of highly dispersed Ru nanoparticles supported on carbon. According to Figure 3.11, (Ru)HKUST-1 transforms in an inert environment; according to thermogravimetric and X-ray diffraction results, CO_2 is released because of decarboxylation of Hbtc and H_2pydc phenyl rings when temperature is raised above 250 °C. Benzylic rings polymerize upon departure of CO_2 with Ru nanoparticles positioned in the polymeric matrix. The pyrolysis eventually leaves an amorphous carbonaceous matrix, which will act as the support for Ru. Under this temperature-programmed treatment, mixed-valence Ru ($Ru_2^{II,III}$) is reduced to

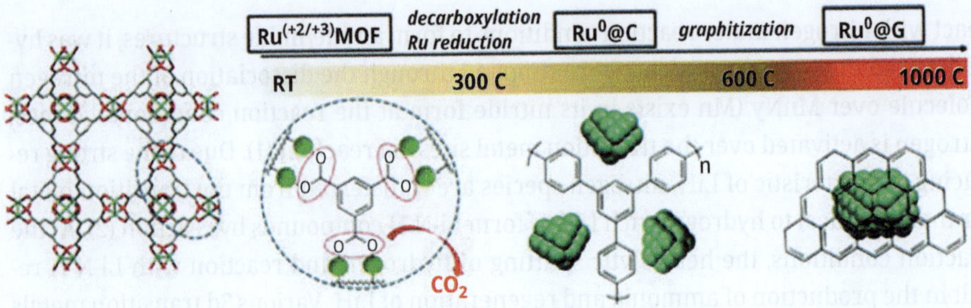

Figure 3.11: Transformation of defected RuMOF via temperature program graphitization. Reprinted from Igancio Luz et al., MOF-derived nanostructured catalysts for low-temperature ammonia synthesis, Catalysis Science & Technology, Volume 10, Issue 1, pp. 105–112, 2020 with permission from Royal Society of Chemistry.

Ru^0 in the presence of carbonaceous material that can act as reducing agents [196]. To complete the graphitization of the nanostructured Ru, the temperature was increased to above 600 °C and up to 1000 °C. Ru nanoparticles sinter into larger clusters while still being highly dispersed in the carbonaceous support. Extending the thermal treatment (i. e., above 12 h) leads to the formation of highly ordered stacked graphitic carbon, but with lower carbon content and surface area.

To further increase the dispersity of Ru while reducing crystal size, nitrogen-doped defected Ru(N)MOFs were prepared using (Ru)HKUST-1 containing equimolar organic ligands of Hbtc and pyridine-3,5-dicarboxylic acid (H_2pydc). The catalyst was washed and thermally treated by the same procedure. Compared to RuMOF, surface areas exceeding 500 m^2/g and Ru nanocrystals of 2–3 nm were obtained for Ru(N)MOF treated at 500 °C. Further increase of the heat treatment up to 900 °C (which is required for the polymerization and graphitization of benzylic rings) resulted in significant collapse of the catalyst structure (200 m^2/g). Ru(N)MOF was found to be not as active as RuMOF, even though the nanostructured Ru was more dispersed.

RuMOF was then treated with acetylene at 500 °C to make precise nanostructures of core/shell metal/C. The treatment helped stabilize Ru when it was pyrolyzed to temperatures as high as 900 °C. CO chemisorption tests (Figure 3.12a) and TEM images (Figure 3.12b) confirmed the promise of using acetylene to uncover highly dispersed Ru nanoparticles in RuMOF treated with acetylene. As shown in Figure 3.12c, the activation energy of the acetylene-treated RuMOF was found to be 73.74 kJ/mol versus RuMOF with 108 kJ/mol. In general, their activity tests at temperatures in the range of 300–375 °C suggest that acetylene treatment can enhance the catalytic rates by a factor of 2.5 at lower temperature, which can be attributed to improved nanostructured Ru dispersed over the carbonaceous support.

Parvathikar et al. later scaled up the catalyst synthesis of nanostructured Ru over MOFs at 1 g and 200 g batch scales via single-step solvothermal synthesis, as schematically shown in Figure 3.13 [300]. Through this work, the authors further simplified Ru-

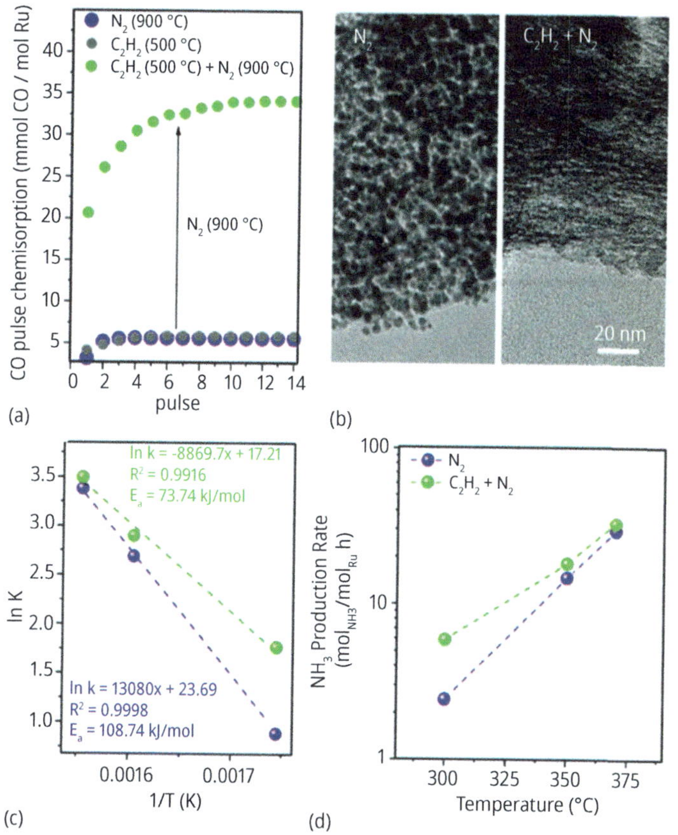

Figure 3.12: (a) CO chemisorption of RuMOF graphitized with N_2 up to 900 °C, C_2H_2 up to 500 °C, and with C_2H_2 up to 500 °C then N_2 to 900 °C. (b) TEM images of N_2- and C_2H_2-treated RuMOF catalysts showing the improved dispersion of Ru with C_2H_2 treatment. (c) Comparison of the activation of RuMOF treated with N_2 and C_2H_2. (d) Rates of ammonia synthesis with RuMOF N_2- and C_2H_2-treated RuMOF catalysts. Adapted from Igancio Luz et al., MOF-derived nanostructured catalysts for low-temperature ammonia synthesis, Catalysis Science & Technology, Volume 10, Issue 1, pp. 105–112, 2020 with permission from Royal Society of Chemistry.

MOF synthesis in a single step, followed by pyrolysis, as illustrated in Figure 3.13a. They showed that nanostructured Ru over C could be obtained in a single step solvothermal process by mixing ruthenium(III) chloride ($RuCl_3$), water, acetic acid, Hbtc, and graphitic carbon in a 25 L reactor and heated to 160 °C for 24 h. The resulting catalyst (Ru/C) was washed and evacuated. Compared to the two-step process, the single-step process does not require the use of acetic acid, acetic anhydride, methanol, and lithium chloride—obviating the need for downstream separation and cleaning. In this effort, various carbon sources with different textural properties (BET 18–1400 m^2/g) were evaluated in the single-step solvothermal process. The resulting catalysts were also exposed to harsh conditions such as treatment with air at elevated temperatures followed by hydrogen

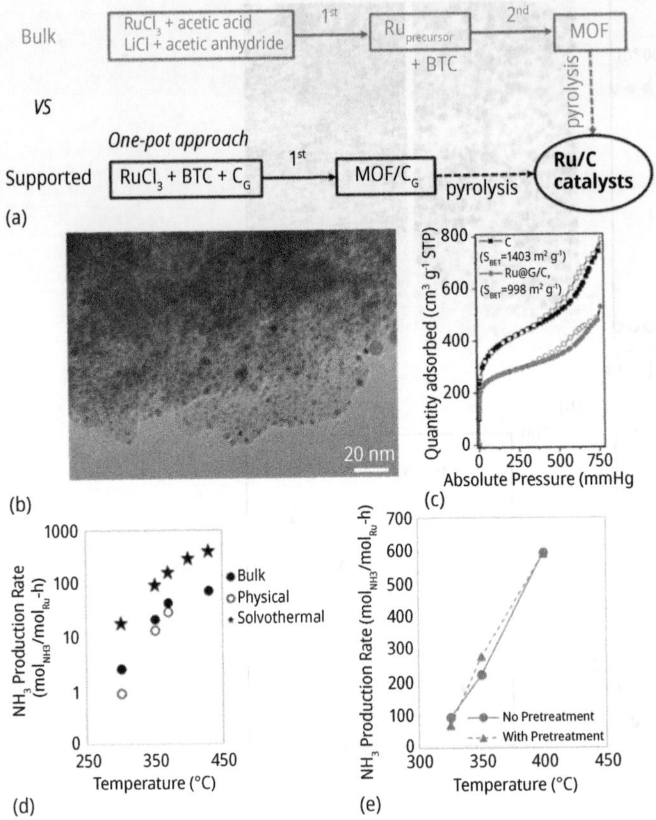

Figure 3.13: (a) Two pathways to synthesize RuC catalysts, one through a two-step process and the other through single step, both followed by pyrolysis. (b) TEM image of RuMOF produced through single-step one-pot approach. (c) Nitrogen sorption isotherms of carbon support and Ru@G/C. (d) Ammonia synthesis rates of with different catalysts at 95 bar and 60,000 h^{-1}. (e) Effect of pretreatment on the activity of catalysts made with Aldrich carbon. Reaction tests were conducted at 95 bar and 60,000 h^{-1}. Adapted from Catalysis Today, Volume 387, Sameer Parvathikar et al., Solvothermal synthesis of MOF-derived supported Ru nanocatalysts for low-temperature ammonia synthesis, pp. 23–27, Copyright 2022, with permission from Elsevier.

reduction at 900 °C to evaluate the robustness of their performance. These are critical tests for catalysts, which are expected to be used in processes exposed to intermittency of renewable energy sources and, therefore, the possibility of changing operating conditions. As shown in Figure 3.13b, TEM images display highly dispersed Ru nanoparticles on carbon with an average particle size of 2 nm. BET displayed the largest surface around 1,000 m^2/g with a bimodal porous structure (Figure 3.13c). Different batches of catalyst production on the pilot scale resulted in catalysts with similar activities, indicating the reproducibility of the proposed single-step solvothermal process for the production of highly dispersed nano-Ru on carbon, as shown in Figure 3.13d. In addition, the catalyst produced with the single-step solvothermal process showed competitive activity compared to catalysts synthesized through other methods, in the temperature range of 300–440 °C. More importantly, the catalyst exposed to hot air and reduced at 900 °C did not show any sign of loss of activity (Figure 3.13e).

4 Improving separation for ammonia production

This chapter discusses the state-of-the-art separation technologies that can be used in the Haber--Bosch process, as well as viable approaches that can further improve ammonia separation in the synthesis loop. Such improvements can come in the form of more complete ammonia separation, lower synthesis loop pressure requirement, and/or separation at a higher temperature. Specific alternative separation technologies discussed in this chapter include condensation, absorption into solids or liquids, adsorption, and membranes.

4.1 Separation in the Haber–Bosch synthesis loop

Assuming that the Haber–Bosch reactor effluent is at more than 550° and 150 bar, the single-pass nitrogen conversion through the reactor is approximately 17.5 %, which is equivalent to 9.7 % mol fraction of ammonia, as illustrated in Figure 4.1. Thus, most of the reactants should be separated and recycled back to the reactor. The conventional Haber–Bosch synthesis loop utilizes phase-change condensation separation, which imposes several limitations to equilibrium-limited reactions, particularly those in which the product does not condense at above-ambient conditions. Overall, the Haber–Bosch synthesis loop requires a relatively large temperature swing, recycle compression, and refrigeration (for synthesis loops operating at lower pressure). As pressure decreases, the capital costs associated with the Haber–Bosch synthesis loop decrease. However, the operating costs associated with separation increase as more refrigeration work is needed. Additionally, the single-pass conversion decreases, indicating a lower overall yield. Hence, a larger portion of the reactants goes through the temperature swings and recycle compression to produce ammonia. Therefore, a path to improve ammonia production utilizing renewable resources is to improve ammonia separation through selective removal of ammonia from the synthesis loop.

As discussed in the previous chapter, over a century of extensive research has been conducted on improving the ammonia synthesis reaction. Comparatively, the research on ammonia separation has been much more limited, with most of the efforts being pursued in the last two to three decades. The key challenge of developing an ammonia separation that fits the Haber–Bosch synthesis loop is the need to operate at elevated pressure and temperature (if high-temperature separation is pursued). With the advancement in materials design and discovery, many more opportunities have been created and opened for the development of novel ammonia separation processes. In this chapter, a critical summary of various separation technologies for the separation of ammonia from the synthesis gas is provided. The separation processes considered fall into three main categories: condensation, sorption-based separations, and membrane-based separations.

https://doi.org/10.1515/9783110762525-004

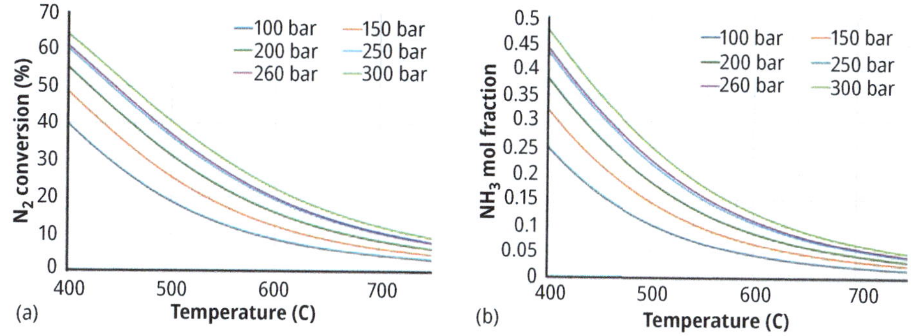

Figure 4.1: (a) Nitrogen conversion and (b) mole fraction of ammonia in a single pass at different temperatures and pressures. A higher pressure is in favor of greater conversion while a higher temperature is in favor of faster reaction rates.

4.2 Condensation

Phase-change condensation is the most prevalent form of separation practiced in the chemical and refining industry. A schematic process flow diagram of the Haber–Bosch synthesis loop is shown in Figure 4.2. As shown, the synthesis gas consisting of ammonia (see Figure 4.1 for the composition) exits the converter train (R-101) at temperatures above 550 °C. To condense ammonia, the synthesis gas should pass through a chain of heat exchangers and coolers to reduce the temperature to the ambient temperature. Then the gas mixture enters the condensation unit. For lower pressure processing, refrigeration may be required.

Ammonia condenses at −33.3 °C at a partial pressure of 100 kPa. Increasing the ammonia partial pressure is in favor of condensation at a higher temperature, thus reducing the need for extensive refrigeration of a synthesis gas mixture. This is the key reason why the Haber–Bosch synthesis loop pressure is elevated. Increasing the synthesis loop pressure above 200 bar can obviate the need for refrigeration. This reduced operating cost comes at a significant increase in capital costs (compression, cost of materials, etc.). This shows the significant role operating pressure plays in the balance of operating costs versus capital costs. It is predicted that roughly 12 % of the energy consumption in the synthesis loop is attributed to refrigeration and precooling of the synthesis gas (total energy consumption being 9.7 kWh/kg_{NH3}) [223]. Although an extensive amount of energy is consumed for cooling and refrigeration, the separation remains incomplete with 2–3 % of ammonia escaping the separation unit and returning to the reactor inlet, which thermodynamically is a major source of exergy destruction [55].

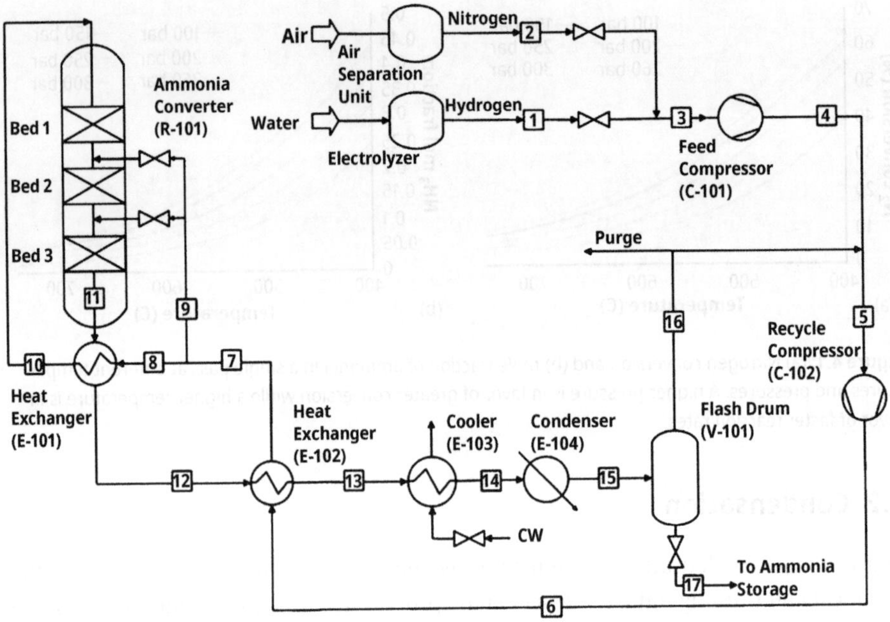

Figure 4.2: Process flow diagram of conventional Haber–Bosch synthesis loop. The hydrogen-nitrogen mixture goes through the feed compressor, gets mixed with the recycled gas, and then goes through the recycle compressor. The temperature of synthesis gas increases via heat integration with reactor effluent stream. Then the synthesis gas is further heated and then introduced to the converter chain, which can be operated in various modes (cold shot mode displayed here). The effluent gas mixture is then cooled through a chain of heat exchanger before being introduced to refrigerated condensation (typically at −20 °C) to separate ammonia. Reprinted with permission from Bosong Lin et al., Performance of a Small-Scale Haber Process: A Techno-Economic Analysis, ACS Sustainable Chemistry & Engineering, Volume 8, Issue 41, pp. 15517–15531. Copyright 2020 American Chemical Society.

4.3 Sorption-based separations

Sorption-based separations used for ammonia separation fall into two main categories: adsorption and absorption. In the gas adsorption process, the desired product (ammonia) adheres to the surface of the solid material because of the weak intermolecular forces such as van der Waals forces. Solid adsorbents are typically highly porous and have a large surface area. The key separation mechanism in adsorbents for ammonia separation from the synthesis gas is size exclusion. The main adsorbents studied for ammonia separation are zeolites, silica, activated carbon, porous organic polymers, and metal organic frameworks. In contrast, ammonia absorption is based on the reaction of ammonia with the adsorbent via the formation of a chemical bond, which results in high selectivity for ammonia. Ammonia absorption with conventional solvents, ionic liquids, deep eutectic solvents, and solid absorbents are the main materials studied to date.

4.3.1 Solid adsorbents

4.3.1.1 Conventional adsorbents

One of the early efforts to evaluate the sorption of ammonia in various organic and inorganic adsorbents is by Helminen et al. [134]. They evaluated the performance of a variety of zeolites (including 4A, 5A, 13X, dealuminated faujasite, dealuminated pentasil, and clinoptilolite) and organic adsorbents (including activated carbon, charcoal, and sulfonated and nonsulfonated polymeric sorbents). Amberlyst 15 showed the highest ammonia capacity at higher ammonia partial pressure (195 mg/g), while at lower partial pressures, zeolites showed better results in efficiently capturing ammonia. As shown in Figure 4.3, the sharp ammonia uptake at a lower ammonia partial pressure indicates the superior performance of these materials (except for Purolite MN200) for ammonia separation from gas mixtures with a low ammonia partial pressure. In contrast, these sharp isotherms suggest insufficient regeneration of the adsorbents upon pressure swing. Among the adsorbents tested, dealuminated zeolites, nonsulfonated polymeric sorbents, and activated carbon displayed less favorability for ammonia adsorption.

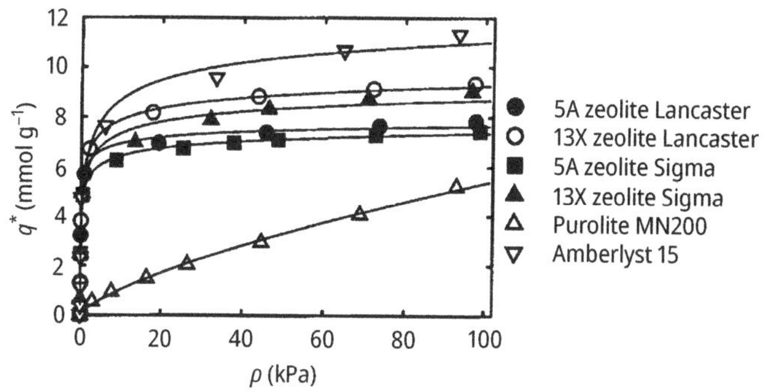

Figure 4.3: Isotherms of ammonia adsorption at 25 °C. Figure displays experimental data points and solid lines representing Toth isotherms fitted to the experimental results. Reprinted with permission from Jarkko Helminen et al., Adsorption equilibria of ammonia gas on inorganic and organic sorbents at 298.15 K, Journal of Chemical & Engineering Data, Volume 46, Issue 2, pp. 391–399.. Copyright 2001 American Chemical Society.

To further investigate the potential of using adsorbents for the separation of ammonia at higher temperatures and evaluate the possibility of pressure swing for regeneration, Helminen et al. measured the isotherms of 13X zeolite, 4A zeolite, alumina, silica gel, and activated carbon at 25 and 120 °C and ammonia partial pressures of 1–100 kPa [133]. They defined a parameter called working capacity, which is the difference between the equilibrium capacity at 1 and 100 kPa partial pressure of ammonia. They found that activated carbon showed the highest working capacity of 93.5 $g_{NH3}/g_{sorbent}$ at 25 °C, while zeolites (i. e., 4A and 13 X) displayed almost similar working capacities of 56 $mg_{NH3}/g_{sorbent}$, regardless of the temperature range studied. The promise of Y-type zeolite absorbents for ammonia separation with temperature and pressure swings was further evaluated by Liu and Aika [226]. They measured the adsorption isotherms of various ion-exchanged Y-type zeolite (Na-Y, H-Y, Co-Y, Cu-Y, K-Y, Rb-Y, and Cs-Y). As shown in Figure 4.4, the key finding was that the requirement to have a reasonable working capacity for Y-type zeolites was to swing the temperature from 50 to 200 °C; pressure swing was found to be marginally suitable to obtain reliable working capacities. This finding is widely accepted in other adsorption-based separations, suggesting that the change in temperature improves the regeneration of adsorbents. It is important to note that the working capacities discussed here are obtained from volumetric tests, which is an equilibrium-based experimentation, thus lacking the kinetics involved in determining the actual working capacity.

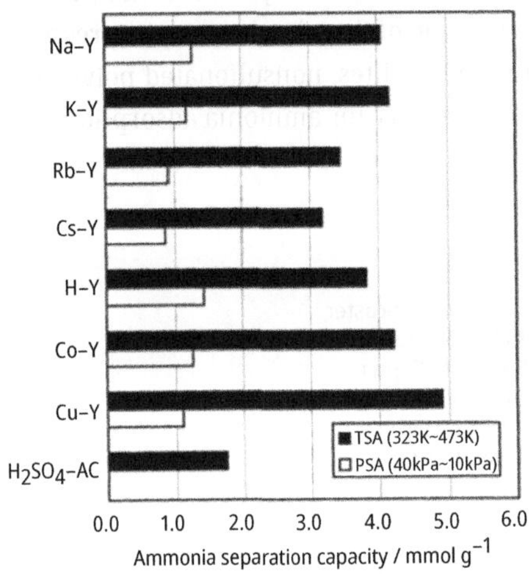

Figure 4.4: Ammonia working capacities of different adsorbents (including ion exchanged Y-type zeolites and activated carbon) evaluated at various pressure and temperature swing conditions. Figure is reprinted from [226] under Creative Commons CC BY license.

Most of the reports discussed previously were focused on static volumetric analysis tests to evaluate the adsorbents. Ouyang et al. evaluated the performance of a variety of organic and inorganic adsorbents in dynamic adsorption breakthrough tests [289]. They studied the separation of ammonia from dilute gaseous streams with the goal of improving the purity of hydrogen streams. The goal was to reduce the ammonia concentration to less than 0.1 ppm to generate a stream of fuel cell-quality hydrogen from ammonia cracking. As can be seen in Figure 4.5, some of the adsorbents (i. e., silica gel, activated carbon, 3A zeolite, and 4A zeolite) showed early breakthroughs in the sorption tests, although they demonstrated promising volumetric adsorption capacities in earlier reports. This suggests that probably the kinetics of the ammonia adsorption in these adsorbents do not meet the need for designing a cyclic steady-state process for

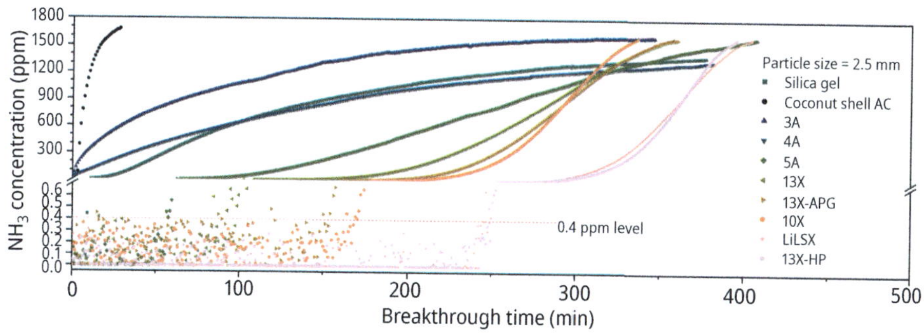

Figure 4.5: Adsorption breakthrough profiles for various adsorbent evaluated at 1.7 % ammonia in nitrogen at 25 °C. Adapted from International Journal of Hydrogen Energy, Volume 46, Issue 64, Wei Ouyang et al., Dynamic ammonia adsorption by FAU zeolites to below 0.1 ppm for hydrogen energy applications, pp. 325590–32569, Copyright 2021, with permission from Elsevier.

ammonia adsorption. In contrast, faujasite zeolites showed sharp breakthroughs. The absorber leaked less than 0.4 ppm ammonia throughout the test. Although such clear breakthroughs showed the promise of faujasite zeolites for ammonia capture, still better separations would be of demand to reach sub-0.1 ppm ammonia concentration in the adsorbent column exhaust.

Carbonaceous materials are among the adsorbents studied for ammonia separation and capture and activated carbon is one of the most studied materials. The surface chemistry of activated carbon was found to have a great influence on its ammonia capacity [111]. By adding oxygenated functional groups via nitric acid treatment, the ammonia capacity at room temperature can improve significantly. Some reports suggested that activated carbon with carboxylic functional groups showed improved ammonia capacity [60]. These findings later motivated Cho et al. to impregnate activated carbon with a variety of organic acids to enhance ammonia separation from gaseous streams with low ammonia partial pressures [60]. Among various adsorbents synthesized, succinic acid-activated carbon (13.3 wt.%) showed the highest ammonia capacity of 200 $mg_{NH3}/g_{sorbent}$ in the isotherm tests at 25 °C and 1 atm. The capacity of the sorbents decreased to only 32 $mg_{NH3}/g_{sorbent}$ when breakthrough tests were carried out with 100 ppm ammonia at 25 °C.

Following the role of surface chemistry in carbonaceous materials, Bandosz et al. reported a series of detailed studies on the use of graphite oxides and the importance of surface chemistry for ammonia separation [333, 334, 304]. In their first effort, graphite oxide was synthesized from commercial graphite using the Hummers method. Samples calcined in nitrogen or dried in air were compared for ammonia adsorption. It was suggested that for the adsorbents that have denser carboxylic groups, ammonia acted as either Bronsted or Lewis acids, while for calcined GO with denser epoxy groups, nucleophilic substitution was suggested as the major mechanism of adsorption. Of importance was the specifics provided on different modes of ammonia adsorbed (strongly vs.

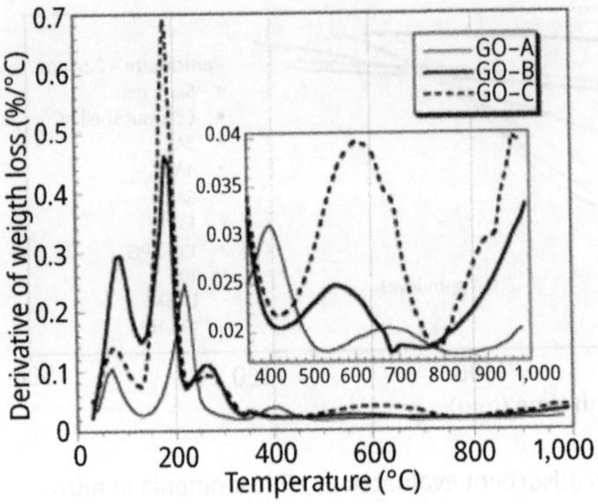

Figure 4.6: Thermo-gravimetric analysis of ammonia release from different graphite oxide adsorbents. Reprinted from Carbon, Volume 47, Issue 2, Mykola Seredych et al., Role of graphite precursor in the performance of graphite oxides as ammonia adsorbents, pp. 445–456, Copyright 2009, with permission from Elsevier.

weakly). According to the thermogravimetric analysis results shown in Figure 4.6, physically adsorbed water was released at around 100 °C for sorption tests in the presence of moisture. The second peak between 150–200 °C belongs to the decomposition of epoxy groups bonded with ammonia and the exfoliation of GO. The peak between 250–450 °C was attributed to the decomposition of carboxylic groups. At temperatures above 500 °C, phenolic/hydroxyl groups and ethers decompose. These findings were supported by potentiometric titration results [334].

4.3.1.2 Porous organic polymers

Porous organic polymers (POPs) are built from organic building blocks linked through covalent bonds. They possess a network of interconnected pores that remain after the removal of any guest molecules used in their synthesis. This permanent porosity is crucial for their applications. POPs often have large surface areas, sometimes exceeding thousands of square meters per gram. This large surface area is a consequence of the porous structure and contributes to their ability to adsorb and interact with various substances. POPs can be functionalized by acidic groups, metal sites, or hydrogen bonds. A summary of POPs reported for ammonia separation provided in Table 4.1 indicates that the ammonia adsorption capacity in this class of materials can reach up to 340 $mg_{NH3}/g_{sorbent}$ when the polymeric network is functionalized with Co(II) [235, 236].

4.3.1.3 Metal organic frameworks

Metal organic frameworks (MOFs) are a class of porous materials composed of metal ions or clusters coordinated to organic ligands, which create highly porous structures with tunable properties. The use of MOFs and their compounds has been investigated for the separation, capture, and storage of ammonia [381, 124]. The major challenge of

Table 4.1: Ammonia adsorption capacity for different porous organic polymers. Reprinted with permission from Jiayin Zhang et al., A mini review on NH_3 separation technologies: recent advances and future directions, Energy & Fuels, Volume 36, Number 24, pp. 14516–14533. Copyright 2022 American Chemical Society. To find the references for this table, readers should refer to [419].

POP	Functional groups	Adsorption conditions	NH_3 capacity (mmol/g)
BPP-5	acidic	25 °C, 100 kPA	17.70
P_1-PO_3H_2	acidic	25 °C, 100 kPA	18.70
PAA	acidic+hydrogen bond	25 °C, 100 kPA	10.70
1TCS	acidic	25 °C, 100 kPA	8.52
1S	acidic	25 °C, 0.05 kPA	4.03
PAA	acidic+hydrogen bond	25 °C, 9 kPA	5.20
MPI-A-30	hydrogen bond	30 °C, 100 kPA	12.41
PIP-X	metal	25 °C, 100 kPA	10.20
PIM-1-COOH	acidic	25 °C, 100 kPA	12.20
MPOP-1.0-SO_3H	acidic	25 °C, 100 kPA	10.96
PDVB-2.0AA	acidic	25 °C, 5 kPA	3.53
$(SrOOC)_{17-}$	metal	25 °C, 100 kPA	14.30
Ph2Im-TBBTA	metal	25 °C, 100 kPA	20.10
P4VP-R8-Co	metal	25 °C, 100 kPA	20.10
PVIm-Pyd	metal	25 °C, 100 kPA	22.01

MOFs for ammonia applications is designing materials that are stable and can provide reproducible capacities for long periods. The proposed approaches to improve MOF resistance in the presence of ammonia include the use of inertial metal ions and increasing the heterolytic metal–ligand bond strength [311]. Materials of Institute Lavoisier (MIL) are among the most popular groups of MOFs that have been tested for ammonia separation, including MIL-53, NH2-MIL-53, MIL-100, and MIL-101 with capacities of 75, 92, 136, and 170 $mg_{NH3}/g_{sorbent}$ at 25 °C and 100 kPa. The stability of these MOFs was evaluated only for five cycles and in moderate conditions [56]. The ammonia sorption capacity can be further intensified by incorporating open metal sites (e. g., HKUST-1 and MOF-74), although open sites will deteriorate the stability of MOFs in the presence of ammonia [182]. The critical role of the open metal site on the enhanced ammonia uptake was demonstrated on UiO-66, in which the hydroxyl sites were decorated with Cu(II) sites [241]. Incorporation of Cu(II) sites resulted in an almost two-fold increase in UiO-66 uptake. In-situ diffraction and spectroscopic results indicate that this enhancement should be attributed to Cu(II)-NH_3 interaction. The results indicate the need for more energy-intensive regeneration with open metal sites, which is not in favor of the energy efficiency of the Haber–Bosch synthesis loop [187].

Two different tunable frameworks used for ammonia separation, capture, and storage are M2(dobpdc) (M= Mg^{2+}, Mn^{2+}, Co^{2+}, Ni^{2+}, Zn^{2+}; $dobpdc^{4-}$ = 4,4'-dioxidobiphenyl-3,3'-dicarboxylate) and M(cyhdc) (M: Cu; $cyhdc^{2-}$ = trans-1,4-cyclohexanedicarboxylate), first reported by Kim et al. [187] and Long et al. [350]. Both MOFs demonstrated excellent stability and high surface areas. For the former, the organic linker ($dobpdc^{2-}$) is the dianionic form of a biphenyl derivative, where two benzene rings are connected through a single bond, with two carboxylate groups ($-COO^-$) and two hydroxyl groups (-OH, which are deprotonated as oxido groups, $-O^-$). M2(dobpdc) forms hexagonal one-dimensional channels, as shown in Figure 4.7a. The open metal sites (Mg^{2+}) coordinated to the $dobpdc^{2-}$ linker were found to be responsible for the unprecedented ammonia capacities observed (23.9 $mmol_{NH3}/g_{sorbent}$ at 100 kPa and 25 °C), as illustrated in Figure 4.7. Kim et al. studied the regeneration of M2(dobpdc) frameworks under vacuum and 250 °C for 24 hours. Even with such extensive desorption conditions, two of the frameworks (Mg and Ni) displayed a decreased capacity in three consecutive cycles. M2(dobpdc) demonstrated a reasonable ammonia uptake of 8.37 $mmol_{NH3}/g_{sorbent}$ through fixed-bed breakthrough experiments with 1000 ppm

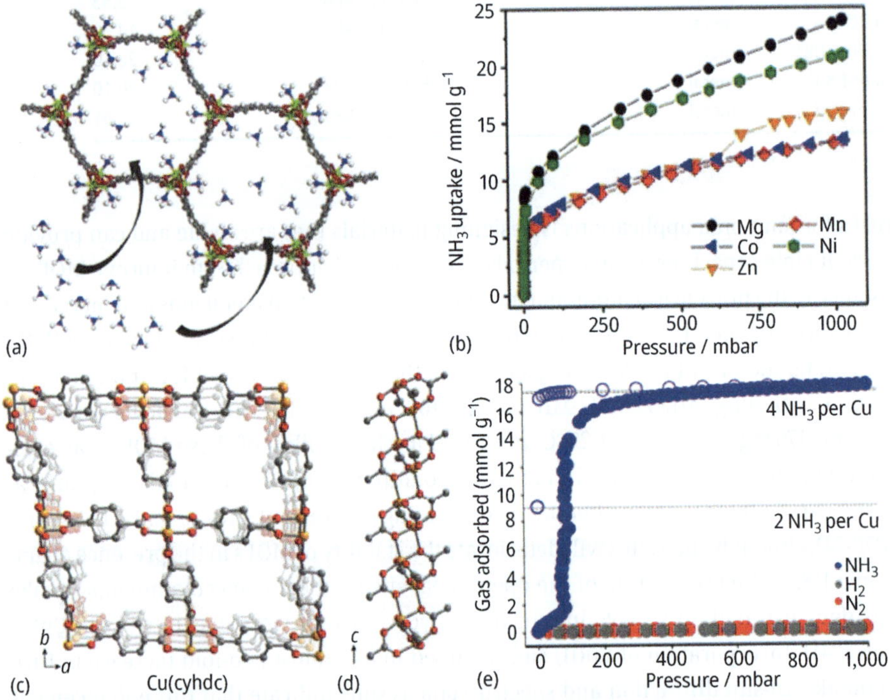

Figure 4.7: (a) Hexagonal channels of M2(dobpdc). (b) Isotherms of ammonia adsorption on M2(dobpdc) ($M^{2+} = Mg^{2+}$, Mn^{2+}, Co^{2+}, Ni^{2+}, and Zn^{2+}) at 25 °C. (c) and (d) structure and side view of one-dimensional Cu(cyhdc). (e) ammonia adsorption isotherm of Cu(cyhdc) at 25 °C. Adapted from Benjamin E. R. Snyder et al., A ligand insertion mechanism for cooperative NH_3 capture in metal–organic frameworks, Nature, Volume 613, pp. 287–291, 2023 with permission from Springer Nature and from Adapted from Dae Won Kim et al., High Ammonia Uptake of a Metal–Organic Framework Adsorbent in a Wide Pressure Range, Angewandte Chemie International Edition, Volume 59, pp. 22531–22536, 2020 with permission from John Wiley and Sons.

ammonia in the feed stream at 100 kPa with the balance helium. Inspired by Cu(bdc) MOF made of copper (Cu^{2+}) ions and 1,4-benzenedicarboxylate (bdc^{2-}) as the organic linker, Snyder et al. developed a crystalline air-stable framework Cu(cyhdc) ($cyhdc^{2-}$ = trans-1,4-cyclohexanedicarboxylate), in which copper(II) sites were coordinatively saturated [350]. The $cyhdc^{2-}$ ligand features a cyclohexane ring with carboxylate groups at the 1 and 4 positions in a trans configuration, providing a flexible yet structured framework. They demonstrated selective reversible ammonia adsorption owing to the selectivity of the framework for ammonia over nitrogen and hydrogen. Cu(cyhdc) displayed a stepped ammonia uptake isotherm, which can be leveraged for efficient pressure swing adsorption schemes to facilitate the regeneration of the sorbents. The framework displayed 318 $mg_{NH3}/g_{sorbent}$ at 100 kPa and 0 °C. No breakthrough results have been reported on this framework.

In general, MOFs are interesting candidates for ammonia separation, capture, and storage. Kim et al. provided a comprehensive summary of the ammonia storage capacity in a variety of MOFs, as well as comparison with other supports (see supporting information in [187]). Opportunities in ammonia separation include designing MOFs with improved thermal swing stability, reduced energy input for regeneration, and simplified synthesis procedure at scale [124].

4.3.2 Liquid absorption

4.3.2.1 Conventional solvents

Absorption of gases in liquid solvents has been widely used in a variety of chemical processes. Liquid absorption is commonly used in various industrial applications, such as gas scrubbing [291], air pollution control [185], and chemical manufacturing [135], the most prominent application being CO_2 capture technologies [10]. When a gas comes into contact with a liquid solvent, the molecules of the gas interact with the molecules of the liquid. The solvent selectively absorbs certain gases on the basis of their solubility in the liquid. The extent to which the gas is absorbed into the liquid depends on the solubility of the gas in the solvent. Thus, a key factor in the selection of the solvent is the solubility of the desired gaseous compound in the solvent. Water can dissolve ammonia (530 g/L at 20 °C) but the key shortcoming is the instability of ammonia within the water. With a slight increase in temperature, a significant amount of ammonia can be released. That is why acidic pH is used for more effective ammonia capture [16]. Some of the solvents that have high ammonia solubility are polar solvents [343], glycerol [82], and polyethylene glycol [418]. Another shortcoming with using conventional solvents for ammonia separation in the Haber–Bosch synthesis loop is their high partial vapor pressure, which results in significant solvent loss and possible contamination. The catalyst is also sensitive to any impurities; therefore, extensive scrubbing is required before recycling the gas mixture to the reactor inlet. In general, conventional solvents do not offer a viable option for ammonia separation in the Haber–Bosch synthesis loop.

4.3.2.2 Ionic liquids

Ionic liquids (ILs) are a class of compounds that are composed entirely of ions and charged particles and are in a liquid state at or near room temperature. Typically, they are salts made up of large organic cations (positively charged ions) and smaller anions (negatively charged ions). Unlike most traditional salts, which have high melting points and are solid at room temperature, ILs remain in liquid form because their ionic components are large and bulky, which prevents them from packing closely together, thus lowering the melting point. One of the most important features of ILs is their extremely low vapor pressure, which makes them safer and more environmentally friendly compared to conventional solvents. ILs have been widely used for ammonia capture and separation, primarily for ammonia removal from air streams. Few of the previous reports target ammonia separation and capture in the Haber–Bosch synthesis loop. ILs were first put into test for ammonia capture by Yokozeki et al., where they studied the solubility of ammonia in conventional ILs, including 1-butyl-3-methylimidazolium hexafluorophosphate ([bmim][PF6]), 1-hexyl-3-methylimidazolium chloride ([hmim][Cl]), 1-ethyl-3-methylimidazolium bis(trifluoromethylsulfonyl)imide ([emim][Tf2N]), and 1-butyl-3-methylimidazolium tetrafluoroborate ([bmim]-[BF4]) [412]. Later, Shi and Maginn used Monte Carlo simulations of the osmotic ensemble to compute ammonia absorption isotherms in [emim] [Tf2N], and their results showed strong agreement with previous experimental measurements [340]. The calculated enthalpy mixing for ammonia in the ionic liquid ranges from −2 to −11 kJ/mol, with the magnitude increasing as the ammonia concentration increases and the temperature decreases. Ammonia is found to interact more strongly with the [emim] cation than the [Tf2N] anion, due to the formation of a strong hydrogen bond between the basic nitrogen in ammonia and the acidic ring hydrogen atoms on the cation. This contrasts with observations for other solutes like carbon dioxide and sulfur dioxide in ionic liquids. Since then, the use of ILs for ammonia separation and capture has been widely expanded and progressed.

Using a hybrid computational-experimental approach, Palomar et al. studied 272 different **conventional ILs** (different cations imidazolium, pyridinium, pyrrolodinium, quinolinium, ammonium, phosphonium, thioronium, [choline$^+$], [N-EtOH$^+$], and [EtOHmim$^+$]), and anions ([FEP$^-$], [NTf_2^-], [$FeCl_4^-$], [$CF_3SO_3^-$], [$CF_3CO_2^-$], [$B(CN)_4^-$], [NO_3^-], [$CH_3CO_3^-$], [$CH_3SO_3^-$], [$CH_3SO_4^-$], [$C(CN)_3^-$], [DCN$^-$], [BF_4^-], [SCN$^-$], [PF_6^-], [CHOO$^-$], and [Cl$^-$]) to discover ILs with higher capacity for ammonia [293]. It was found that in order to tune the solubility of ammonia in ILs, it is critical to optimize the hydrogen bond donation of the cation, rather than changing the anion. ILs with ammonium and imidazolium were found to be more favorable for increasing the ammonia absorption capacity. In terms of anions, it was found that highly fluorinated anions were more in favor of a higher ammonia capacity. An overview of their findings is illustrated in Figure 4.8.

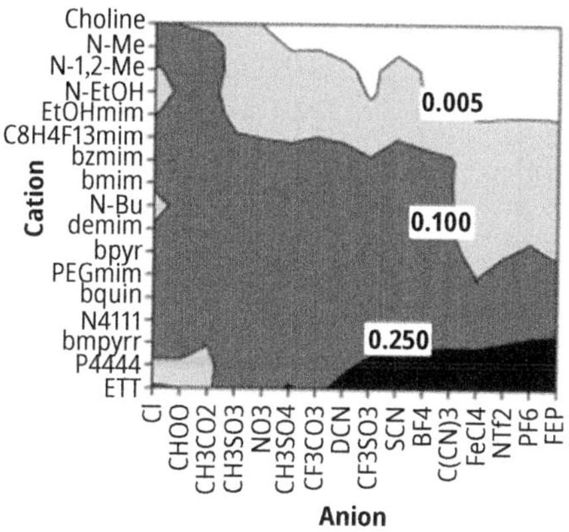

Figure 4.8: Overview of Henry's law constants for 272 Ionic Liquids studied by Palomar et al. for ammonia at 25 °C [293]. Reprinted from Separation and Purification Technology, Volume 82, Jose Palomar, Task-specific ionic liquids for efficient ammonia absorption, pp. 43–52, Copyright 2011, with permission from Elsevier.

Various forms of ILs have been investigated for ammonia separation. **Hydroxyl ILs**—ionic liquids that contain a hydroxyl group (-OH) as part of the cation or anion—were among the first candidates to improve the solubility of ammonia in ILs. Palomar et al. found that functionalization of ILs with hydroxyl groups (e. g., [choline$^+$]) significantly improved IL capacity for ammonia [293]. Larrechi et al. also studied the solubility of ammonia in three hydroxylic ILs and found that at lower concentrations of ammonia, hydroxyl IL capacities exceeded those of chollinium ILs [206]. Li et al. also reported the same observation that hydroxyl-functionalized imidazolium displayed higher ammonia absorption capacities due to stronger hydrogen bonding between ammonia and the hydrogen atom of the hydroxyl group [216]. Ammonia capacities of up to 71 mg_{NH3}/g_{IL} were obtained with 1-(2-hydroxyethyl)-3-methylimidazolium tetrafluoroborate [EtOHmim][BF_4] at 40 °C and 100 kPa. A more comprehensive report evaluated the effect of the anions, as illustrated in Figure 4.9a, and it was found that the ammonia solubility was in the following order for different anions: $[NTf_2]^- > [PF_6]^- > [BF_4]^- > [SCN]^- > [DCA]^- > [NO_3]^-$. **Protic ILs**—ILs with cations derived from protonated organic compounds and the anions typically being inorganic or organic anion have also been shown to be promising candidates for ammonia capture and separation. Shang et al. found that the ammonia capacity in 1-butylimidazolium bis(trifluoromethylsulfonyl)imide ([Bim][NTf2]) was 10 times higher than conventional [Bmim][NTf2] [337]. Later, they evaluated more protic ILs and, interestingly, they discovered that the choice of anion (i. e., the acidity of the anions) has a greater influence on the solubility of ammonia in protic ILs, as illustrated in Figure 4.9b. Capacities as high as 113 mg_{NH3}/g_{IL} were obtained at 40 °C and 100 kPa. **Metallic ILs**—cation is a metal complex/ion, as opposed to the usual organic cations—have been also found to be promising candidates for ammonia dissolution. Chen et al. reported the ammonia capacity in the synthesized

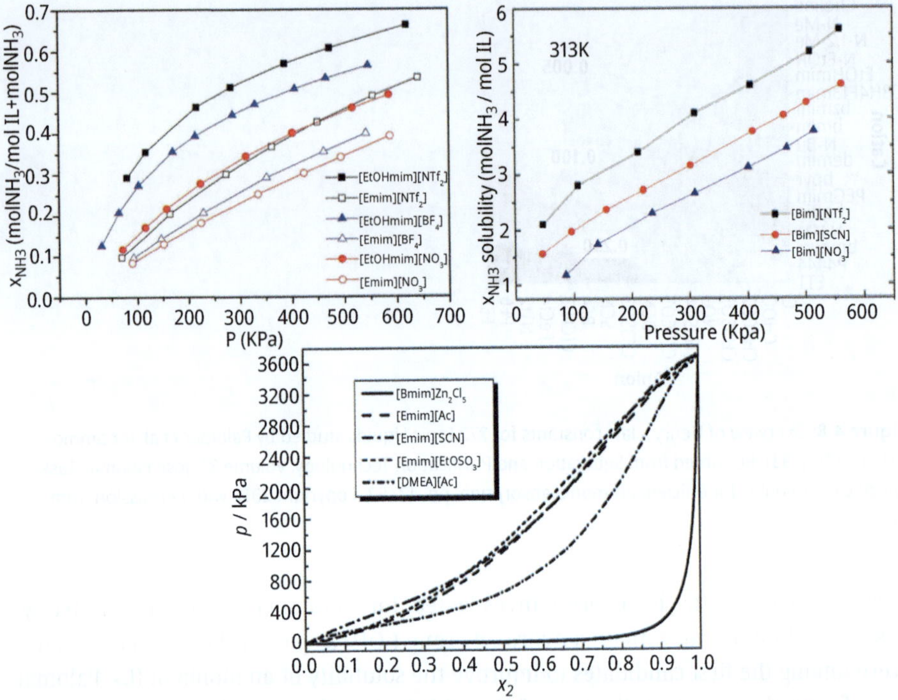

Figure 4.9: Examples of ammonia solubility in (a) hydroxyl functionalized, (b) protic, and (c) metalic ILs. (a) solubility of ammonia in [EtOHmim]$^+$-based ILs with six different anions (40 °C). (b) solubility of ammonia in protic ILs with three different anions (40 °C). (c) Ammonia purification with different metallic ILs (75 °C). Adapted with permission from Lu Zhang et al., An overview of ammonia separation by ionic liquids, Industrial & Engineering Chemistry Research, Volume 60, Issue 19, pp. 6908–6924. Copyright 2021 American Chemical Society.

1-butyl-3-methylimidazolium zinc pentachloride [Bmim] [ZnitrogenCl$_5$], along with four other ILs [56]. An ammonia capacity as high as 488 mg$_{NH3}$/g$_{IL}$ was obtained at 50 °C and 210 kPa. Various metallic ILs with metals including copper, tin, nickel, manganese, cadmium, zinc, and iron were evaluated [392]. Apparently, complexation of the Zn metal site with ammonia led to this superior performance for the separation of ammonia. **ILs with multiple functional sites** are also pursued, in which a combination of functional groups is incorporated into ILs. The combination of functional groups is expected to enhance the ammonia capacity. For example, in one of the first studies, (DPIL) [CnOHim] X (n = 1, 2; X = [NTf$_2$]$^-$, [BF$_4$]$^-$, [SCN]$^-$) was synthesized by introducing weak acidity protons and hydroxyl groups into an imidazolium ring [414]. Quantum chemical calculations suggest that both the weak acidity proton (primary) and the hydroxyl group (secondary) act as ammonia binding sites. The summary of the capacities of various types of ILs for the separation and storage of ammonia is provided in Tables 4.2 to 4.6, obtained from Zhang et al. [420].

Table 4.2: Ammonia adsorption capacity for different conventional ionic liquids at relevant properties. Adapted with permission from Lu Zhang et al., An overview of ammonia separation by ionic liquids, Industrial & Engineering Chemistry Research, Volume 60, Issue 19, pp. 6908–6924. Copyright 2021 American Chemical Society. To find the references for this table, readers should refer to [420].

Ionic liquid	Temperature (°C)	Pressure (kPa)	mol_{NH3}/mol_{IL}	g_{NH3}/g_{IL}
[Emim][BF_4]	25	140	0.282	0.024
[Hmim][BF_4]	25	220	0.485	0.032
[Omim][BF_4]	25	120	0.389	0.023
[Bmim][BF_4]	40	180	0.250	0.019
[Bmim][PF_6]	25	174	0.541	0.032
[Bmim][SCN]	30	145	0.32	0.028
[Bmim][DCA]	30	114	0.30	0.025
[Bmim][NTf_2]	40	101	0.28	0.011
[Bmmim][DCA]	30	115	0.25	0.019
[Bmmim][SCN]	30	96	0.18	0.015
[Bmmim][NTf_2]	40	100.49	0.20	0.008
[Hmim][CI]	25.1	111	0.326	0.027
[Hmim][SCN]	30	101	0.20	0.015
[Emim][NTf_2]	25.4	145	0.159	0.007
[Emim][Ac]	25.2	463	1.506	0.151
[Emim][$EtOSO_3$]	25.1	421	1.075	0.077
[Emim][SCN]	25.1	307	0.799	0.080
[C_2Py][NTf_2]	40	101	0.20	0.009

Table 4.3: Ammonia adsorption capacity for different protic ionic liquids at relevant properties. Adapted with permission from Lu Zhang et al., An overview of ammonia separation by ionic liquids, Industrial & Engineering Chemistry Research, Volume 60, Issue 19, pp. 6908–6924. Copyright 2021 American Chemical Society. To find the references for this table, readers should refer to [420].

Ionic liquid	Temperature (°C)	Pressure (kPa)	mol_{NH3}/mol_{IL}	g_{NH3}/g_{IL}
[TMGH][BF_4]	20	100	1.08	0.090
[TMGH][NTf_2]	20	100	1.17	0.050
[Eim] [NTf_2]	40	234	3.66	0.165
[Eim][DEP]	40	101	0.460	0.031
[Mim][NTf_2]	40	268	2.79	0.131
[Mmim][NTf_2]	40	209	3.26	0.147
[Bim][SCN]	40	158	2.36	0.219
[Bim][NO_3]	40	236	2.29	0.208
[Bim][NTf_2]	40	101	2.69	0.113
[2-mPy][NTf_2]	40	101	3.078	0.140
[2,6-2mPy][NTf_2]	40	101	3.049	0.134
[Py][NTf_2]	40	101	2.946	0.139

Table 4.4: Ammonia adsorption capacity for different hydroxyl-functional ionic liquids at relevant properties. Adapted with permission from Lu Zhang et al., An overview of ammonia separation by ionic liquids, Industrial & Engineering Chemistry Research, Volume 60, Issue 19, pp. 6908–6924. Copyright 2021 American Chemical Society. To find the references for this table, readers should refer to [420].

Ionic liquid	Temperature (°C)	Pressure (kPa)	mol_{NH3}/mol_{IL}	g_{NH3}/g_{IL}
[Choline][NTf_2]	20	100	1.857	0.082
[MTEOA][$MeOSO_3$]	20	101	3.545	0.232
[DMEA][Ac]	25.1	278	1.604	0.183
[$TMGHPO_2I$][BF_4]	20	100	0.71	0.038
[EtOHmim][NTf_2]	40	128	0.830	0.035
[EtOHmim][PF_6]	40	108	0.674	0.042
[EtOHmim][BF_4]	40	116	0.659	0.052
[EtOHmim][SCN]	40	95	0.311	0.029
[EtOHmim][NO_3]	40	103	0.253	0.023
[EtOHmim][DCA]	40	140	0.360	0.032
[$HOOC(CH_2)_3mim$][NTf_2]	40	101	1.54	0.058

Table 4.5: Ammonia adsorption capacity for different metallic ionic liquids at relevant properties. Adapted with permission from Lu Zhang et al., An overview of ammonia separation by ionic liquids, Industrial & Engineering Chemistry Research, Volume 60, Issue 19, pp. 6908–6924. Copyright 2021 American Chemical Society. To find the references for this table, readers should refer to [420].

Ionic liquid	Temperature (°C)	Pressure (kPa)	mol_{NH3}/mol_{IL}	g_{NH3}/g_{IL}
[Bmim][Zn_2Cl_5]	50	103.5	8.025	0.305
$[Emim]_2[Co_2(NCS)_4]$	30	101	5.99	0.178
$[Bmim]_2[Co_2(NCS)_4]$	30	101	6.03	0.163
$[Hmim]_2[Co_2(NCS)_4]$	30	101	6.09	0.151
$[Bmim]_2[Li(BF_4)_3]$	40	101	3.819	0.119

Table 4.6: Ammonia adsorption capacity for different multiple site ionic liquids at relevant properties. Adapted with permission from [420]. Copyright 2021 American Chemical Society. To find the references for this table, readers should refer to [420].

Ionic liquid	Temperature (°C)	Pressure (kPa)	mol_{NH3}/mol_{IL}	g_{NH3}/g_{IL}
[EtOHim][NTf_2]	40	100	3.11	0.135
[EtOHim][BF_4]	40	100	2.47	0.210
[EtOHim][SCN]	40	100	2.23	0.222
[MeOHim][NTf_2]	40	100	3.04	0.136
[2-Mim][NTf_2]	40	101	3.037	0.142
[Im][NTf_2]	40	101	3.461	0.169
[EtOHpi][NTf_2]	40	101	2.406	0.100
[Piper][NTf_2]	40	101	1.771	0.082
[Eim][$Li(NTf_2)_2$]	40	101	6.618	0.169
[2-Mim][$Li(NTf_2)_2$]	40	101	7.012	0.183

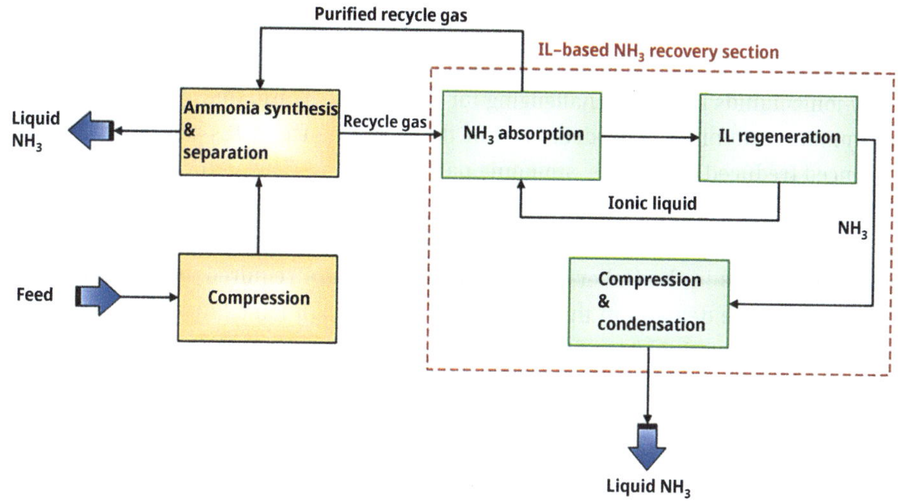

Figure 4.10: Proposed modified Haber–Bosch synthesis loop with IL-based separation for ammonia recovery from the reactor effluent. Reprinted from Separation and Purification Technology, Volume 301, Sensen Shi et al., Simulation and energetic assessment of the ammonia synthesis loop with ionic liquid-based ammonia recovery from recycle gas, Article 121951, Copyright 2022, with permission from Elsevier.

Shi et al. recently published one of the most relevant reports with respect to use of ILs for ammonia separation [339]. They reported the simulation and energetic assessment of Haber–Bosch synthesis loop modified with IL based ammonia recovery, as displayed in Figure 4.10. They selected protic IL [Bim][NTf2] as their sorbent for ammonia recovery. Two different process configurations were proposed, one with three flash tanks vs. two flash tanks plus a stripping column. Overall, results suggest that the synthesis loop with stripping column displayed the best performance in terms of energy intensity, although it resulted in higher capital investment. The lowest specific energy consumption (0.98 MJ/mt_{NH3}) was achieved by optimizing the ammonia content in the recycle gas and synthesis reactor inlet.

ILs have been considered as promising sorptive materials for the selective separation of ammonia in the Haber–Bosch synthesis loop because of their unique properties, such as tunable solubility, low vapor pressure, and thermal stability. However, there are yet several limitations associated with the use of ILs, which creates new opportunities for further research. The most critical challenge to be addressed is designing ILs with lower viscosity and less energy-intensive regeneration. The high viscosity of the ILs considered for the Haber–Bosch synthesis loop has remained a challenge to circumvent; recirculating ILs within the process can become challenging and costly [263]. Additionally, a higher viscosity limits the mass transfer rates for the gas sorption process, thus reducing the efficiency of gas separation. The energy required for regeneration may not always be economically or thermodynamically favorable, especially if the ionic liquid's absorption capacity is not sufficiently high. Second, ILs that are cheaper and easier for bulk synthesis are of interest. The complexity of scaled processes for the synthesis of ILs can lead to high costs for large-scale applications. Scaling up the production of high-purity ionic liquids has been challenging for this reason. Third, tuning ILs to improve the ammonia solubility is needed. Most of the performance of ILs can be drastically influenced (reduced) at a lower ammonia partial pressure. This reduces the chance of operating an IL-based Haber–Bosch synthesis loop at low pressure. ILs can face challenges when exposed to high temperatures in ammonia synthesis. Further studies on the stability of ILs in the Haber–Bosch synthesis loop are required, while decreasing the heat exchange demand in the synthesis loop. Although ionic liquids are generally considered more environmentally friendly than organic solvents because of their low vapor pressure and nonvolatility, some ionic liquids contain components that could be toxic or environmentally harmful. The toxicity of ionic liquids depends on their cations and anions, and improper disposal or leaks could pose environmental risks.

In summary, while ionic liquids show great potential for ammonia separation in the Haber–Bosch synthesis loop, challenges related to viscosity, cost, scalability, thermal stability, environmental impact, and solvent regeneration must be addressed to make them commercially viable for large-scale applications. The IL section of this chapter was inspired and organized based on a review article by Zhang et al. [420].

4.3.3 Deep eutectic solvents

Deep Eutectic Solvents (DES) are formed by mixing two or more components, typically a hydrogen bond donor (HBD) and a hydrogen bond acceptor (HBA), in specific molar ratios. When combined, these components form an eutectic mixture, which has a lower melting point than either of the individual components. It was first DES of blended 1-butyl-3-methylimidazolium mesylate ([Bmim][MeSO3]) and urea that were evaluated for ammonia sorption. Later, a variety of other reports studied DES for ammonia capture, separation, and storage [78, 394, 428, 58, 427, 217, 83, 122, 77, 76]. Imidazolium hydrogen chloride ([ImH]Cl) and glycerol (Gly)—a weak acid site and strong hydrogen-bond donor site—were used to develop DES and later evaluated for ammonia separation. This DES was very attractive for sorbing ammonia at lower partial pressures. An ammonia capacity of 345 mg_{NH3}/g_{IL} is reported at 25 °C and 101 kPa. This high capacity was attributed to the strong acid-base interaction between ([ImH]Cl) and ammonia [240].

4.4 Solid absorption in metal halides

Metal halides have been considered for the storage of ammonia. Probably the first application of metal halides for ammonia separation and storage was for heat pumps and refrigeration with the metal halide-ammonia pair [355]. It was first Aika et al. that considered the separation of ammonia using metal halide salt, with the ultimate purpose of integration into the Haber–Bosch synthesis loop [225]. Metal halides react with ammonia to form stoichiometric ammine complexes following the reaction below:

$$MH(s) + nNH_3(g) \leftrightarrow M(NH_3)_n H_x(s) + \Delta H \tag{4.1}$$

This reaction is typically a multistep process, in which different coordination numbers (n) will be achieved. The coordination number obtained at equilibrium is a function of reaction conditions (reaction and pressure) and the type of metal halide. When an ammonia partial pressure (critical pressure P_{298}) is obtained at 298 K, the transformation of the metal halide and the uptake of ammonia into the crystal site begin. Depending on the pressure, different coordination numbers will be achieved. The coordination numbers and P_{298} listed in Table 4.7 for metal chlorides and Table 4.8 for metal bromides are those that are found to be the most stable structure. The critical pressure can be used indirectly as a measure of how tight the metal halides hold ammonia molecules within the crystal structure. The lower the P_{298}, the stronger the binding and the higher the enthalpy of the formation reaction. The critical pressures listed in Tables 4.7 and 4.8 indicates the strength of bonds formed between ammonia and metal sites within the metal halide crystals. This can be used to guide the design of separation units based on metal halides and to decide on the removal efficiency based on the choice of the metal halide.

The absorption of ammonia into metal halide salts starts with solid gas reaction at the surface of the metal halide. Upon the formation of an ammine complex and the existence of the required driving force (partial pressure of ammonia greater than P_{298}), ammonia starts to diffuse into the salt crystals to form a stable ammine complex. The most stable structures of $MgCl_2$ in different amminated phases are shown in Figure 4.11. Hexa-ammine $MgCl_2$ forms a K_2PtCl_6 type structure, as shown in Figure 4.11b. Mg is body-centered, and ammonia molecules are face-centered in every other cube of chlorine atoms.

Metal halides have also been considered for indirect hydrogen storage or polishing hydrogen streams from ammonia prior to feeding fuel cells, owing to their high capacity and affinity for ammonia. In terms of ammonia storage, Vegge et al. have published a series of reports on tuning the properties of metal halides for ammonia storage and release, with particular attention paid to atomic-scale simulations to understand the structure of metal halides for facile uptake and release [379]. Among the alternative separation methods for ammonia separation in the Haber-Bosch synthesis loop, metal halide absorbents are the most studied to date. The efforts fall into three main categories:

- Metal halide absorbent chemistry: Developing chemistries of metal halides that can be used for reproducible uptake and release of ammonia

Table 4.7: Enthalpies, entropies, and critical pressure P_{298} for different metal chlorides for different stable coordination. [226].

Sample	ΔH ($J\,mol^{-1}$)	ΔS ($J\,mol^{-1}$ K−1)	P_{298} (kPa)
$ZnCl_2$ 10-6	29,588	219.23	1853.2
$CuCl_2$ 10-6	31,387	227.72	2490.2
$SnCl_2$ 9-4	31,806	224.86	1490.8
$PbCl_2$ 8-3.25	34,317	223.76	474.3
$BaCl_2$ 8-0	37,665	227.25	187.0
$SnCl_2$ 4-2.5	38,920	229.82	153.5
$PbCl_2$ 3.25-2	39,339	230.27	136.8
$CaCl_2$ 8-4	41,013	230.30	69.9
$SrCl_2$ 8-1	41,431	228.80	49.3
$CaCl_2$ 4-2	42,268	229.92	40.3
$ZnCl_2$ 6-4	44,779	230.24	15.2
$SrCl_2$ 1-0 (a)	15,316	131.43	15.2
$PbCl_2$ 2-1.5	46,035	230.89	9.9
$PbCl_2$ 1.5-1	47,290	232.50	7.2
$MnCl_2$ 6-2	47,416	228.31	4.2
$ZnCl_2$ 4-2	49,467	230.24	2.3
$CuCl_2$ 6-3.3	50,241	230.75	1.8
$FeCl_2$ 6-2	51,266	227.99	8.46×10^{-1}
$CuCl_2$ 3.3-2	56,497	237.22	3.11×10^{-1}
$CoCl_2$ 6-2	53,986	228.10	2.86×10^{-1}
$PbCl_2$ 1-0	55,660	231.04	2.07×10^{-1}
$MgCl_2$ 6-2	55,660	230.63	1.97×10^{-1}
$NiCl_2$ 6-2	59,217	227.75	3.33×10^{-2}
$CaCl_2$ 2-1	63,193	237.34	2.12×10^{-2}
$CaCl_2$ 1-0	69,052	234.14	1.36×10^{-3}
$MnCl_2$ 2-1	71,019	232.35	4.95×10^{-4}
$MgCl_2$ 2-1	74,911	230.30	8.04×10^{-5}
$FeCl_2$ 2-1	76,167	231.91	5.88×10^{-5}
$CoCl_2$ 2-1	78,134	232.17	2.74×10^{-5}
$NiCl_2$ 2-1	79,515	232.17	1.57×10^{-5}
$ZnCl_2$ 2-1	80,352	229.72	8.35×10^{-6}
$MnCl_2$ 1-0	84,202	233.18	2.68×10^{-6}
$FeCl_2$ 1-0	86,880	233.01	8.91×10^{-7}
$MgCl_2$ 1-0	87,048	230.88	6.45×10^{-7}
$CoCl_2$ 1-0	88,303	232.80	4.90×10^{-7}
$NiCl_2$ 1-0	89,810	233.01	2.73×10^{-7}
$ZnCl_2$ 1-0	104,625	227.79	3.70×10^{-10}

- Absorptive-enhanced Haber–Bosch: Developing intensified configurations of ammonia production processes
- Ammonia uptake and release with metal halides: Understanding the absorption-desorption behavior in metal halide chemistries

Table 4.8: Enthalpies, entropies, and critical pressure P_{298} for different metal bromides for different stable coordination [226].

Sample	ΔH (J mol^{-1})	ΔS (J mol^{-1} K−1)	P_{298} (kPa)
$CaBr_2$ 8-6	41,647	232.72	72.4
$CaBr_2$ 6-2	45,574	221.09	3.67
$CaBr_2$ 2-1	79,037	253.82	2.58×10^{-4}
$CaBr_2$ 1-0	71,318	226.22	2.10×10^{-4}
$SrBr_2$ 8-2	54,714	259.62	9.45
$SrBr_2$ 2-1	88,706	340.78	1.82×10^{-1}
$SrBr_2$ 1-0	36,259	165.01	1.85×10^{-1}

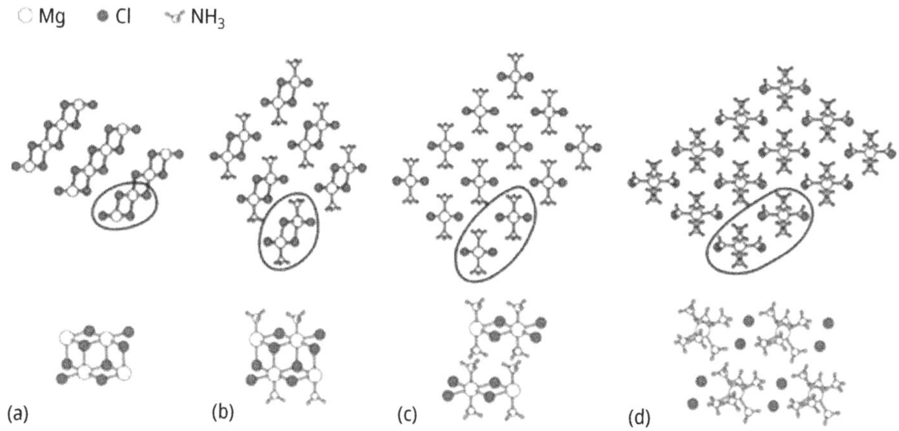

Figure 4.11: Density functional theory predictions of stable structures for (a) $MgCl_2$, (b) $Mg(NH_3)Cl_2$, (c) $Mg(NH_3)_2Cl_2$, and (d) $Mg(NH_3)_6Cl_2$. Adapted with permission from Rasmus Z. Sorensen et al., Indirect, reversible high-density hydrogen storage in compact metal ammine salts, Journal of the American Chemical Society, Volume 130, Issue 27, pp. 8660–8668. Copyright 2008 American Chemical Society.

4.4.1 Metal halide absorbent chemistry

Metal halides were first proposed for ammonia separation in the Haber–Bosch synthesis loop by Aika et al. [225, 228, 227]. In the search for an ammonia scrubber from the reactor effluent, they studied the use of an alkaline metal halide absorbent. The metal halides were pelletized, crushed, then pretreated (evacuated) at various temperatures (25–600 °C) before performing isotherms. They used a physisorption machine to generate isotherm data for different salts evacuated at different temperatures. The key findings of their results are: 1) The evacuation temperature has direct influence on the capacity of the sorbent. Metal halides that hold water molecules tighter within the structure (e. g., $MgCl_2$) require a higher evacuation temperature to dry the metal halides. 2) Large hysteresis trends were observed in the absorption–desorption curves for the majority of the metal halides, and this hysteresis was more pronounced for the absorbents that were evacuated at a lower temperature. 3) None of the salts reached their thermodynamic coordination number. 4) The second uptake was very different from the first uptake and typically less than the first uptake, indicating that ammonia absorption was not 100 % reversible under the desorption conditions used. These findings highlight several points. First, there is a great possibility that some of the salts will decompose in the presence of water molecules. Hydrated metal halides must be dried in a controlled environment. Second, the physisorption machine has some limitations in terms of dwelling time (the time required to reach equilibrium for each data point). Most probably, the reported results are not for a fully equilibrated gas-absorbent system. Third, the physisorption machine is limited to subatmospheric pressure of ammonia, which might not be sufficient to saturate the metal halides, even if the critical pressure is less than 100 kPa. In a different effort, Liu decided to adjust the performance of these metal halides by mixing a variety of metal halides, and results similar to their prior report were observed [225]. The key finding was that alloying the metal halides can tune the sorption and isotherm energies of the absorbent.

Later, a team from the Technical University of Denmark reported fairly detailed computational-experimental studies for designing metal halides for ammonia storage. Essentially, they proposed that hydrogen can be indirectly stored as ammonia in the solid state form [351], and proposed this route as an alternative to the hydrogen economy [61] for ease of storage and transportation, due to its inherently safer conditions. As illustrated in Figure 4.12, the storage densities of 10 kg of hydrogen in $MgCl_2$ and $CaCl_2$ are greater than 100 g/L. Their initial efforts aimed to study the complexation of $MgCl_2$ to form the ammine structure [169, 359]. One of the prominent findings was that metal halides were not mechanically stable. The formation of nanopores upon uptake and release of ammonia was observed. This was attributed to the expansion and shrinkage of the crystallite in these polycrystalline materials, leaving the pore behind after ammonia desorption [359, 145]. Later, their efforts were more focused on $SrCl_2$, predom-

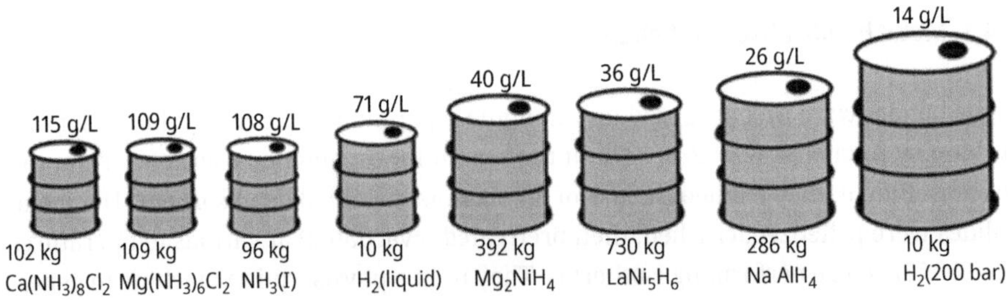

Figure 4.12: Mass and volume of 10 kg hydrogen stored reversibly by 8 different methods, based on the best obtained reversible densities reported in the literature without considering the space or weight of the container. Reprinted from Asjborn Klerke et al., Ammonia for hydrogen storage: challenges and opportunities, Journal of Materials Chemistry, Volume 59, pp. 22531–22536, 2023 with permission from Royal Society of Chemistry.

inantly due to the lower temperature required for the release of ammonia in a more reversible manner [174]. By alloying Sr and Ba, they could reach unprecedentedly high ammonia capacities higher than the capacities of pure metal halides, as summarized in Table 4.9. They used a genetic algorithm and Density Functional Theory (DFT) calculations to predict alloying halide mixtures, with the goal of optimizing absorption energies while maximizing capacity [172]. The developed metal halides were considered primarily for safe storage and release for NO_x removal for selective catalytic reaction. Apparently, the instability of absorbents and heat management resulting from reaction exotherms/endotherms were key obstacles limiting the successful transition of this technology.

Table 4.9: Ammonia capacities (total and usable) reported for various forms of metal chloride ammonia storage. Reprinted from Journal of Solid State Chemistry, Volume 221, Agata Bialy et al., Solid solution barium–strontium chlorides with tunable ammonia desorption properties and superior storage capacity, pp. 32–36, Copyright 2015, with permission from Elsevier.

Metal chloride	g_{NH3}/L_{max}	$g_{NH3(useable)}/L$	g_{NH3}/kg_{max}	$g_{NH3(useable)}/kg$
Anhydrous liquid NH_3	610	610	1000	1000
$Ba_{0.375}Sr_{0.625}Cl_2$	613	604	435	428
$Ba_{0.4}Sr_{0.6}Cl_2$	613	604	433	427
$Ba_{0.5}Sr_{0.5}Cl_2$	605	599	426	422
$BaCl_2$	597	597	396	396
$SrCl_2$	601	526	462	404
$CaCl_2$	670	502	551	413
$FeCl_2$	678	452	446	298
$NiCl_2$	657	438	441	294
$MgCl_2$	639	426	518	345
$MnCl_2$	618	412	448	299
$ZnCl_2$	608	406	429	286
LiCl	550	367	547	364

Later, efforts to directly use metal halide absorbents for the Haber–Bosch synthesis loop were continued at the University of Minnesota. The hypothesis was that metal halides can be used for high-temperature separation of ammonia from the reactor effluent gas mixture, because of the formation of strong chemical bonds between ammonia and metal sites in the ammine structure. Huberty et al. the absorption behavior of ammonia in metal halides at elevated temperatures [144]. They reported ammonia diffusion coefficients in $MgCl_2$ sorbents on the order of 10^{-11} to 10^{-13} cm^2/s. It is important to note that most of the reports to date have studied ammonia absorption on metal halides in static (volumetric) mode and at room temperatures. The delivery of high breakthrough capacities was also critical, as simple static analysis misses the kinetic component in the evaluation of the metal halides for ammonia separation. Wagner et al.

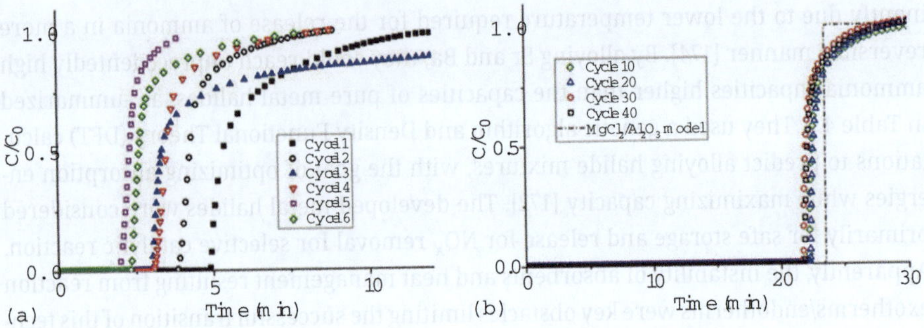

Figure 4.13: Dynamic capacities of ammonia absorption on (a) pure $MgCl_2$ and (b) alumina-supported $MgCl_2$ (6 wt.% loading) absorbents at 150 °C and 100 kPa (N_2:NH_3 = 5:1). Adapted from Kevin Wagner et al., Column absorption for reproducible cyclic separation in small scale ammonia synthesis, AIChE Journal, Volume 63, Issue 7, pp. 3058–3068, 2017 with permission from John Wiley and Sons.

provided one of the first reports on the stability and capacity of metal halides with dynamic (breakthrough) absorption tests [385]. As shown in Figure 4.13a, two of the main challenges with metal halides in breakthrough tests were the instability of metal halides in their pure form and their inability to reach close to their thermodynamic capacity. It was soon found that when metal halides were used in their pure form, the ammonia absorption capaciies in the dynamic breakthrough tests was unstable through multiple cycles and far from their thermodynamic equilibrium capacity (less than 1 %). It was hypothesized that the crystals of metal halides fuse into each other and form a bulky structure, significantly reducing the mass transfer area and decreasing the porosity of the absorbent within the bed. This is intensified with significant volumetric expansion of salt crystals during the ammonia uptake. To avoid these shortcomings, Wagner et al. utilized an inert porous support (alumina) to encage metal halide crystals. As shown in Figure 4.13b, this idea worked to a great extent. Reproducible capacities (beyond 40 cycles) were achieved, while the overall dynamic capacity also improved per mole of salt.

Although supporting metal halides on alumina worked, the overall bed storage densities for ammonia were still relatively low and far from the thermodynamic capacities of the metal halides. For this reason, Malmali et al. took a more unconventional approach [245]. The wet impregnation method was used to have relatively high salt loadings of metal halides supported on mesoporous silica. Silica was chosen in particular because of its larger mesoporous structure and lower costs. For the choice of metal halides, they considered both chloride and bromides of Mg, Ca, and Sr. Bromides can form stronger chemical bonds with ammonia, so they are expected to absorb ammonia more efficiently at elevated temperatures. As can be seen in Figure 4.14a, $CaBr_2$ loadings in the range of 0–100 wt.% were studied. At 60 wt.% mass loading, capacities as high as 85 $mg_{NH3}/g_{sorbent}$ were achieved at 150 °C and 4 bar pressure (N_2:NH_3 = 5:1). Assuming that Si does not contribute to ammonia capacities at such elevated temperatures, the ammonia adsorption capacities per mole of different salts at different temperatures are

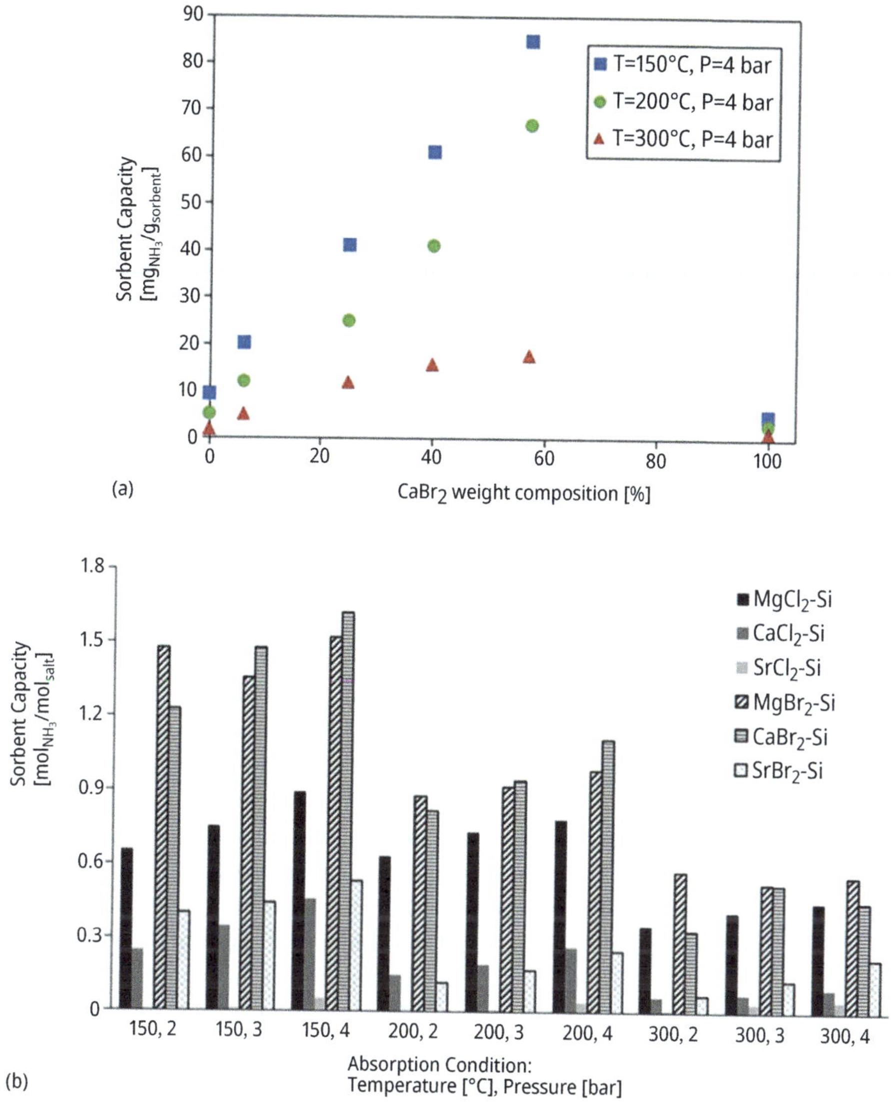

Figure 4.14: (a) High-temperature ammonia absorption working capacities from dynamic breakthrough tests for $CaBr_2$ absorbents supported on silica at 4 bar pressure (N_2:NH_3 = 5:1) for different wt.% loadings, temperatures. (b) Dynamic working capacities of various metal chlorides and bromides supported on silica (40 wt.%) at different temperatures and pressures (N_2:NH_3 = 5:1). Adapted with permission from Mahdi Malmali et al., Better Absorbents for Ammonia Separation, ACS Sustainable Chemistry & Engineering, Volume 6, Issue 5, pp. 6536–6546. Copyright 2018 American Chemical Society.

shown in Figure 4.14b. Different chemistries developed for metal halides display different capacities under target absorption conditions (temperature and ammonia partial pressure). Sharp breakthroughs obtained for silica-supported metal halides (compared to the pure form) indicate fast ammonia uptake within the sorbent structure, which has not been reported. Although chlorides are cheaper and are more appropriate options for large-scale applications, bromides generally exhibit higher capacities. The capacity of bromide-supported salts also displayed a greater dependence on the absorption conditions (i. e., temperature).

The stability of the wet impregnated metal halides on silica was later found to be challenged by the detachment of salt crystals on the support surface. Later, the progress on the use of absorbents for ammonia separation and capture was continued at Texas Tech University. The wet impregnation method was later found to be slightly unstable for extended periods of time. This was notified by the gradual increase in the pressure across the bed. Mahtab et al. compared three different impregnation methods, including incipient wet impregnation (IWI), wet impregnation, and in situ impregnation on SBA-15 and silica [243]. Micrographs and long-term breakthrough cyclic absorption–desorption tests suggested that IWI with the ordered structure of SBA-15 as the support resulted in an excellent stability and reproducible performance. Among the impregnation methods tested, IWI demonstrated the most effective enhancement of reproducible ammonia uptake and sorbent stability. The enhanced stability of the supported absorbent using the SBA-15 support via IWI should be attributed to the improved encaging of the salt crystals. There are fewer salt crystals left outside the SBA-15 support impregnated with the IWI method (Figure 4.15a-b), compared to silica impregnated with the wetness impregnation (WI) method (Figure 4.15c-d). The best performing sorbent, IWI-SBA-15-$MgCl_{2[Mg-OH]}$ (80 $mg_{NH3}/g_{sorbent}$ at 300 °C and 45 kPa), retained its capacity for 50 absorption–desorption cycles with minimal pressure build-up across the bed, making it a promising candidate for scalable industrial applications. The novel in situ impregnation method (ISI) also shows promise by eliminating the need for extensive post-synthesis washing, resulting in absorbents with high capacities (48 $mg_{NH3}/g_{sorbent}$ at 300 °C and 45 kPa) and minimal salt loading. The stability of various $MgCl_2$ sorbents supported on silica and SBA-15 using a variety of impregnation techniques is shown in Figure 4.15e-f, again supporting the notion that IWI impregnation on a mesopore support was more effective in producing stable absorbents with reproducible breakthroughs. Later, Rahman et al. used the IWI method to scale the absorbent geometry, scaling from 500 nm SBA-15 to 3–5 mm silica support while retaining more than 90 % of the capacity (unpublished data).

Through experimental findings, it was found that the management of exotherms within the absorber bed is of extreme importance. The silicon-based absorbent has poor heat conductivity. Thus, it is estimated that mass transport in the absorber is markedly restricted by the rise in temperature in the mass transfer front within the absorber bed. An approach is to optimize the energy of sorption is tuning the absorbent chemistry by alloying different metal halides [72]. The other would be to redesign the chemistry of metal halide sorbents with the goal of designing conductive absorbent that can transport the exotherm heat out of the bed, thus reducing the temperature rise in the

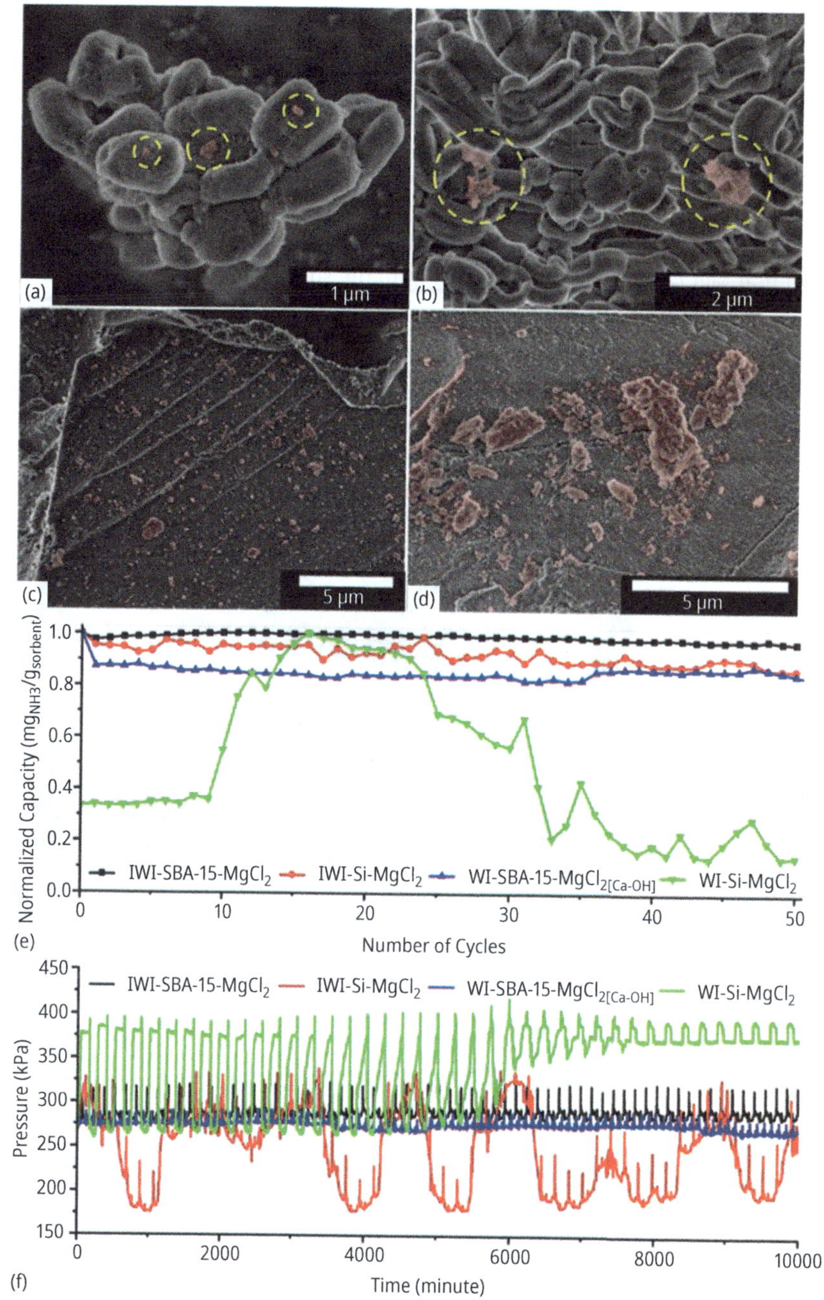

Figure 4.15: Electronic micrographs of MgCl2 absorbents supported on SBA-15 (a)–(b) and Si (c)–(d) using IWI and WI methods, respectively, (e)–(f) Long-term stability of the absorbents evaluated for more than 50 cycles. Adapted from Chemical Engineering Journal, Volume 507, S. M. Tanveer Mahtab and Mahdi Malmali, Insights into structure-composition-capacity relationship of ultrastable metal halide absorbents for ammonia separation and storage, Article 160105, Copyright 2025, with permission from Elsevier.

bed. Nivarty et al. proposed a simple, solventless, and scalable methodology to synthesize metal halide absorbents supported on thermally conductive aluminum fibers [281]. The aluminum fibers were first cut to a submillimeter length. $MgCl_2$ powder was then added to the fibers at 40 wt.% loading, and the mixture was then rotated inside a rock tumbler to further breakdown and mix aluminum fiber and $MgCl_2$. A relatively homogeneous coarse mixture was obtained that was pelletized in a die at 100 MPa. Micrographs of $MgCl_2$ and the aluminum mixture are shown in Figure 4.16a-b. Interestingly, it was found that $MgCl_2$ crystals were adhered onto the surface of the aluminum fiber. This can explain how this physically mixed composition reached a quasi-stable performance. Relatively stable reproducible capacities above 200 $mg_{NH3}/g_{sorbent}$ was obtained at 25 °C and 100 kPa ammonia partial pressure. In general, $MgCl_2$ supported on the metal fiber showed an almost order of magnitude improvement in the thermal conductivity of the absorbent, compared to the baseline Si-supported $MgCl_2$ absorbent. This enhanced thermal conductivity resulted in improved heat responses within the bed, as shown in

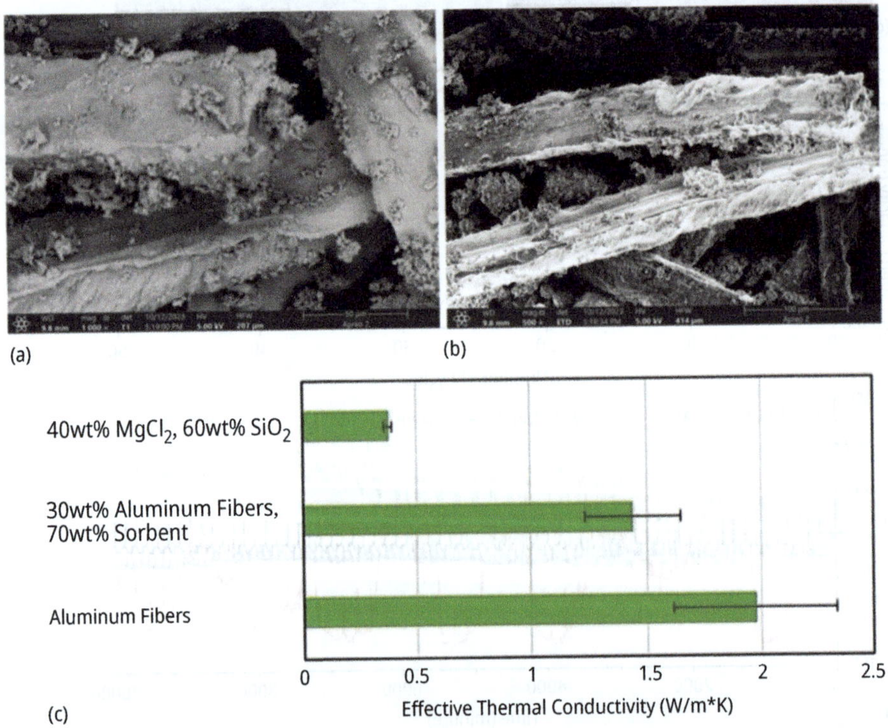

Figure 4.16: (a)–(b) Electronic micrographs of $MgCl_2$ supported on aluminum fibers. (c) Thermal conductivity of the $MgCl_2$ supported on aluminum fibers compare to other. Adapted from Tejas Nivarty et al., Advancing Sustainable Ammonia Production via Solventless, Robust, and Thermally Conductive Absorbents, ACS Sustainable Chemistry & Engineering, Volume 13, Issue 13, pp. 4955–4962. Copyright 2025 American Chemical Society.

Figure 4.16c. Later, Onuoha et al. presented a more detailed study on the composition, stability, and capacity of the $MgCl_2$-aluminium fiber absorbent structure. Compared to other supports (SiO_2, SiC, ethocel, and methocel), the aluminum absorbent with 40 % $MgCl_2$ showed the highest bed thermal conductivity (12 W/m.K) and working capacity (3 mol_{NH3}/mol_{MgCl2}) [286].

4.4.2 Absorptive-enhanced Haber–Bosch

Efforts to design absorptive-enhanced Haber–Bosch were initiated by the report by Himstedt et al., in which they used a fixed bed column packed with $MgCl_2$ to replace condensation in the synthesis loop [136]. Compared to a synthesis loop with only the reactor, it was found that up to 7 times greater conversion was obtained at a sufficiently long time when the absorber bed was included in the synthesis loop. This greater conversion comes at the expense of a 10-fold slower overall production rate because ammonia separation was controlled by the rate of diffusion of ammonia within the absorbents. Considering the fact that ammonia uptake and release in metal halide absorbents were found to be relatively fast, it is probable that the slower rates could be due to other parameters not controlled in the synthesis loop (e. g., recycle flow rate, pressure drop across the bed, etc.). Later, Malmali et al. used the synthesis loop shown in Figure 4.17 to evaluate the performance of ammonia production in a reaction-absorption process

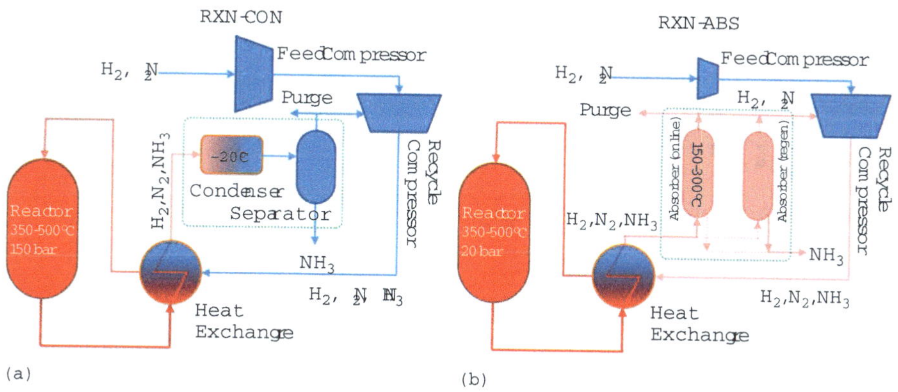

Figure 4.17: Schematic flow diagram for (a) conventional Haber–Bosch synthesis loop operating based on reaction-condensation (RXN-CON) vs. (b) absorptive-enhanced Haber–Bosch synthesis loop operating based on reaction-absorption (RXN-ABS). The reaction section is similar in both processes with reactor feed being heated by its effluent. RXN-CON uses a condenser to liquify product ammonia while RXN-ABS uses a bed a metal halide to remove product ammonia. The RXN-ABS process requires multiple beds to enable cyclic steady-state operation. Adapted with permission from Bosong Lin et al., Performance of a Small-Scale Haber Process: A Techno-Economic Analysis, ACS Sustainable Chemistry & Engineering, Volume 8, Issue 41, pp. 15517–15531. Copyright 2020 American Chemical Society.

[247, 246, 66]. The abbreviation RXN-ABS refers to the Haber–Bosch synthesis loop, in which the condensation is replaced by absorption.

Large granular $CaCl_2$ particles (3–5 mm in diameter) in pure form were used in the absorber packed bed to ensure a uniform distribution of the gas within the bed. Some of the prominent findings are as follows. First, the reaction condensation production rate (RXN-CON conventional Haber–Bosch synthesis loop) at 100 bar was lower than RXN-ABS at 20 bar, with the absorber operating at 180 °C (refer to Figure 4.18a). Second, it was noticed that the absorber bed went through a transition stage; the performance of the RXN-ABS improved in the first five cycles [247]. Third, an unexpected observation was

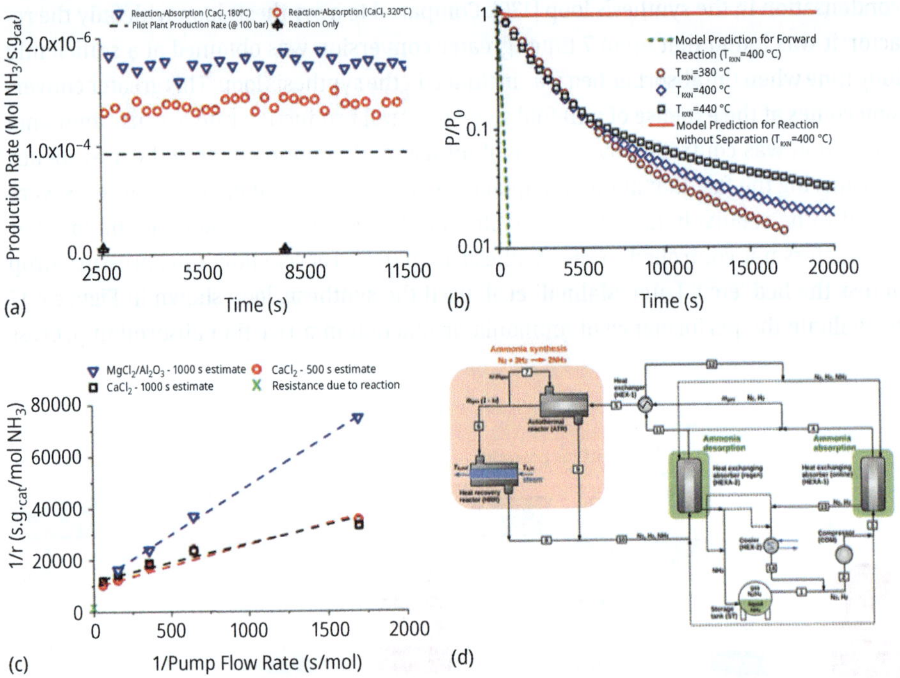

Figure 4.18: (a) Ammonia production rate with absorptive-enhanced Haber–Bosch synthesis loop operating at 20 bar based on RXN-ABS exceeded those of conventional RXN-CON operating at 100 bar. (b) Pressure profiles of batch RXN-ABS experiments starting at 30 bar indicate the RXN-ABS performance depended of recycle flow rate and was independent of reaction temperature at the range studied. (c) Inversed recycle flow rates vs. inverse production rates of RXN-ABS process suggest that recycle flow rate plays an important role in the controlling the RXN-ABS process. (d) Heat recovery cycle for ammonia-based thermochemical energy storage using absorption-based separation. Adapted with permission from Mahdi Malmali et al., Converting Wind Energy to Ammonia at Lower Pressure, ACS Sustainable Chemistry & Engineering, Volume 6, Issue 1, pp. 827–834. Copyright 2018 American Chemical Society and from Oswaldo Andrés-Martínez, Optimal Design of an Absorbent-Enhanced Ammonia Synthesis Process for Solar Thermochemical Energy Storage, ACS Sustainable Chemistry & Engineering, Volume 12, Issue 25, 9446–9460. Copyright 2024 American Chemical Society.

that the RXN-ABS was primarily controlled by how fast the gas was recycled between the reactor and the absorber, not the reactor temperature (Figure 4.18b). A simplified mathematical model predicted an extremum case in which absorption will become limiting at a hypothetical recycle flow rate (Figure 4.18c). Fourth, the process was sustainably producing ammonia at pressures as low as 10 bar. This low pressure approach can provide a sustainable approach for the conversion of renewable energy for distributed chemical manufacturing and energy storage. A recent report proposes the use of RXN-ABS for seasonal energy storage of concentrated solar thermal energy on the utility scale (Figure 4.18d) [20]. Later, Nowrin et al. showed that ammonia production can be further optimized in a cyclic RXN-ABS process and reported production rates as high as 27 $\mu molNH_3/(g_{cats})$ at 30 bar [282].

A comprehensive techno-economic analysis comparing the RXN-CON vs. RXN-ABS synthesis loops was provided by Lin et al. shedding light on the economics and energy efficiency of these two alternatives for green ammonia manufacturing on a scale of 20,000 metric tons per year (mt/y) [223]. The feed compression was found to be the main energy sink in RXN-CON while the recycle compression consumed 74 % of the synthesis loop for recycle compression. An important observation was the effect of the scale of the production for two different synthesis loops, as illustrated in Figure 4.19. When the scale of production decreases, which is closer to what can be expected for distributed manufacturing of chemicals using stranded renewable energy, the RXN-ABS synthesis

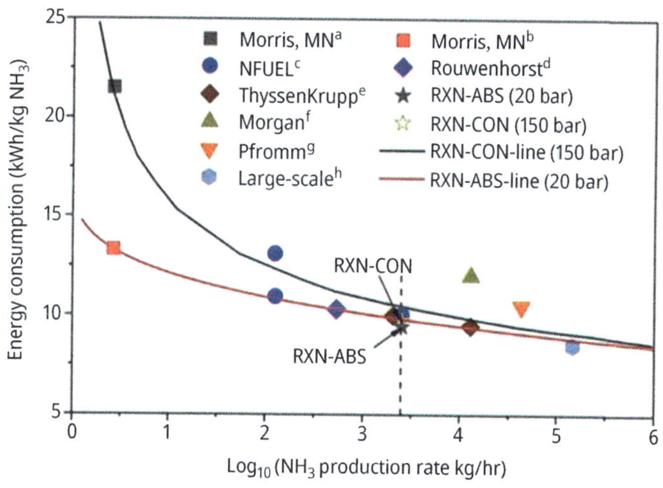

Figure 4.19: Energy consumption in green ammonia production via two different routs (RXN-CON and RXN-ABS) for different scales. At smaller scales, modularity of RXN-ABS is in favor of reducing the energy consumption. Reprinted with permission from Bosong Lin et al., Performance of a Small-Scale Haber Process: A Techno-Economic Analysis, ACS Sustainable Chemistry & Engineering, Volume 8, Issue 41, pp. 15517–15531. Copyright 2020 American Chemical Society.

loop requires less energy. Several cases, including subsidizing green production, learning curves based on the economy of numbers, and obtaining revenue from selling oxygen are considered to make this process feasible. Otherwise, the cost of green hydrogen makes ammonia (\$1,030/mt) almost twice the typical cost of conventional natural gas-based ammonia.

As indicated in the RXN-ABS synthesis loop, the cost of recycle compression can become a limiting factor. An approach to further intensify ammonia production is to combine reaction and separation in one single vessel. Reactants will be compressed into the reactor and ammonia will be produced as a product, as shown in Figure 4.20. The key challenge is the need for high temperature separation that can work at temperatures close to the reaction temperature (above 350 °C). As absorbents have shown promising results for high-temperature ammonia, they open new opportunities for such intensified reactors. Ojha et al. proposed the proof of concept of a single-vacuum reactor housing both the catalyst and absorbents [284]. They used a 1:1 ratio of catalyst to absorbent in their reaction tests. The best results were obtained with $NiCl_2$, in which 33 % pressure decline was observed in the best experiment conducted at 400 °C and 27 bar. Although still at the proof-of-concept level, a significant deviation is observed for an experiment under the same conditions. Additionally, when production rates are calculated, they are far from acceptable rates that can be achieved in the conventional Haber–Bosch synthesis loop. This can be partially attributed to the poisoning of the catalyst with the chlorides. To circumvent this challenge, Smith et al. evaluated a cascaded RXN-ABS unit with the goal of exceeding the equilibrium limitation in the Haber–Bosch converter [348]. The applicability of the proposed reactor design would be questionable as recovering ammonia and heat integration in the proposed design will be a significant challenge.

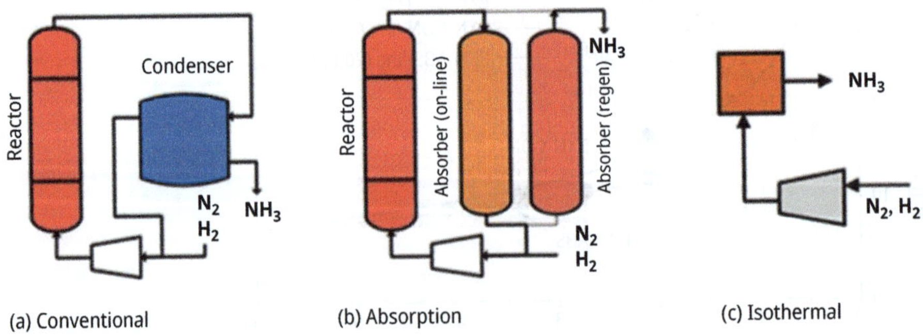

Figure 4.20: Progress toward intensification of distributed ammonia production using absorbent technology. Reprinted with permission from Deepak K. Ojha et al., Integrated Ammonia Synthesis and Separation, ACS Sustainable Chemistry & Engineering, Volume 7, Issue 23, pp. 18785–18792, Copyright 2019 American Chemical Society.

4.4.3 Ammonia uptake and release with metal halides

Ammonia absorption on metal halides was found to be relatively fast, and the regeneration of absorbents requires more energy-intensive processes, including pressure and temperature swings [247, 245]. Kale et al. conducted a systematic analysis examining various pressure and temperature swing conditions, with the goal of determining the working capacity of silica-supported metal halides [177]. Working capacity is defined as the stable capacity of the absorbent achieved after multiple cycles and once the absorbent reaches a reproducible performance. In general, a higher desorption temperature and a lower absorption temperature were in favor of obtaining a greater working capacity. Hrtus et al. also reported optimizing the desorption conditions, with a slight difference: the goal of this study was to alleviate the desorption conditions while increasing the purity of ammonia [142]. This was achieved by controlling the volumetric flow rate of the sweep gas. As shown in Figure 4.21, ammonia purity above 95 % could be achieved when a silica-supported $CaCl_2$ absorbent was operated with the temperature swing between 25 and 300°, with 60 seconds of nitrogen sweep flow at 20 standard cubic centimeters during the desorption. In the majority of reports on ammonia desorption, relatively extensive desorption conditions were considered. Onuoha et al. used more moderate desorption temperatures to partially regenerate the absorbents [285].

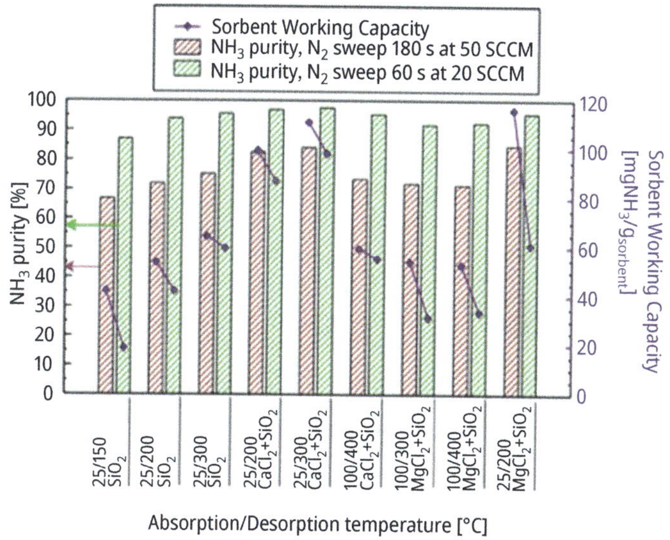

Figure 4.21: Studying the pressure and temperature swings parameters on the regeneration of various metal halide absorbents supported on silica to maximize ammonia purity. Reprinted with permission from Daniel J. Hrtus et al., Achieving +95 % Ammonia Purity by Optimizing the Absorption and Desorption Conditions of Supported Metal Halides, ACS Sustainable Chemistry & Engineering, Volume 10, Issue 1, pp. 204–212, Copyright 2022 American Chemical Society.

4.5 Membranes

Exploiting separations that obviate the need for phase change leads to energy savings of up to one order of magnitude, and membranes are usually considered a promising candidate [342]. Membranes and membrane-based separations generally offer higher energy efficiency, lower energy consumption, simplicity in operation and scale-up, modularity, and most importantly selective molecular separation. Membranes fall into two categories: organic and inorganic. Organic membranes are fabricated from polymeric materials primarily composed of carbon-based materials. Inorganic membranes are made from materials such as metals, oxides, and ceramics. The separation mechanisms in membranes are typically size exclusion, solution-diffusion, charge-charge interactions, or hydrophilic/hydrophobic interactions. For the case of ammonia separation from the synthesis gas, the complexity is raised because of the similarity of the molecules in size. As can be seen in Figure 4.22, the kinematic diameters of hydrogen, nitrogen, and ammonia are very similar in the subnanometer range, making separation based on size exclusion challenging. Thus, hierarchical materials with precise pore structure should be used that allow only the transport of ammonia through the membrane, or additional interactions (i. e., preferred adsorption or diffusion) should be incorporated into the membrane. Another complexity concerns the influence of ammonia on the membrane structure, particularly for polymeric membranes; these membranes typically swell in the presence of ammonia, which compromises the membrane's ability to discriminate between gases over time. Although various transport mechanisms have been proposed, including facilitated transport, ammonia transport appears to be primarily driven by a solution-diffusion process [132, 36, 305]. In the following, we discuss the efforts to design membranes and membrane-based separations for the separation of ammonia from the synthesis gas.

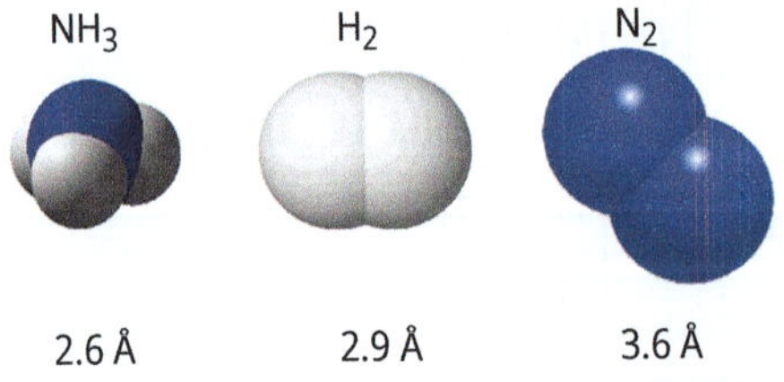

Figure 4.22: Kinematic diameter of ammonia, hydrogen, and nitrogen molecules.

4.5.1 Organic membranes

Membranes fabricated from organic polymers can be synthesized with a wide range of chemical functionalities. This allows the membrane properties to be fine-tuned to specifically target certain molecules for separation. Organic membranes have been considered for gaseous ammonia separation, although there have been challenges that have not yet been resolved. The first report on the use of polymeric films for ammonia separation considered using polyethylene and plasticized trifluoromono-chloroethylene films [48]. They showed that both films show preferential permeation of ammonia over hydrogen and nitrogen. Polymeric films use two mechanisms for the separation of the desired product: molecular sieving and facilitated transport [305]. The efforts in designing organic membranes based on molecular size exclusion have been shown to have little success. Even if partially successful, the membrane performance decreases (particularly for glassy polymers) in the presence of mixture gas due to swelling of the polymer matrix [382, 383]. To resolve this issue in other membrane-based gas separations, chemical cross-linking has been utilized [353, 424] but has not been pursued for membranes that can be applied to gaseous ammonia separation. The results also suggest that efforts to utilize facilitated transport have shown better promise. Polymeric membranes functionalized with the thiocyanate [36] or sulfonate [132, 319, 360] groups displayed intensified ammonia permeabilities. However, this improved membrane performance was only for a single-gas component and declined for the gas mixture. To overcome the problem with decreased performance with a swelled polymeric matrix, Phillip et al. synthesized a bicontinuous triblock copolymer consisting of a continuous rigid polystyrene phase, which provided a solid structure for the membrane to mitigate swelling [305]; a second interpenetrating continuous polystyrene sulfonate phase, which was highly permeable to ammonia.

Supported liquid membranes (SLMs) have been studied for ammonia separation. SLM has two distinct phases: a porous membrane structure that acts as a support and a liquid filler that is impregnated within the pores of the membrane. Essentially, the role of the liquid filler is to facilitate the transport through preferential adsorption or enhanced diffusion of the desired product. For the ammonia separation, the liquid filler can be decided from the solvents that were discussed earlier, either conventional solvents or ionic liquids that are in liquid state at room temperature. Functionalized ILs [2-Mim][NTf_2] and [Im][NTf_2] were impregnated into the pores of a membrane fabricated with sulfonated pentablock copolymer (Nexar^{TM}) [408]. ILs were selected because they show strong interactions with ammonia; this polymeric structure was also selected because of the stability of the membrane and its chain flexibility. [2-Mim][NTf_2] and Nexar^{TM} pair were found to be advantageous and more stable due to the formation of hydrogen bonds between the cation of the IL and the sulfonic acid group of the Nexar^{TM} membrane. The results indicate that the incorporation of ILs in SLM was effective and led to a 4- to 6-fold increase in permeability (see Figure 4.23a) and approximately 3 times improvement in the selectivity of nitrogen (see Figure 4.23b) and hydrogen (see Figure 4.23c) over ammonia. Interestingly, increasing the IL loading in the SLM only slightly improved the SLM performance (permeability and selectivity). It is important to note that IL/Nexar^{TM} SLM was evaluated at ambient temperature and 100 kPa pressure. It is also not clear whether the membrane was tested with the mixture gas (including both hydrogen and nitrogen with ammonia). In general, IL-incorporated SLMs displayed interesting results for ammonia separation. Their performance exceeded those of conventional polymeric

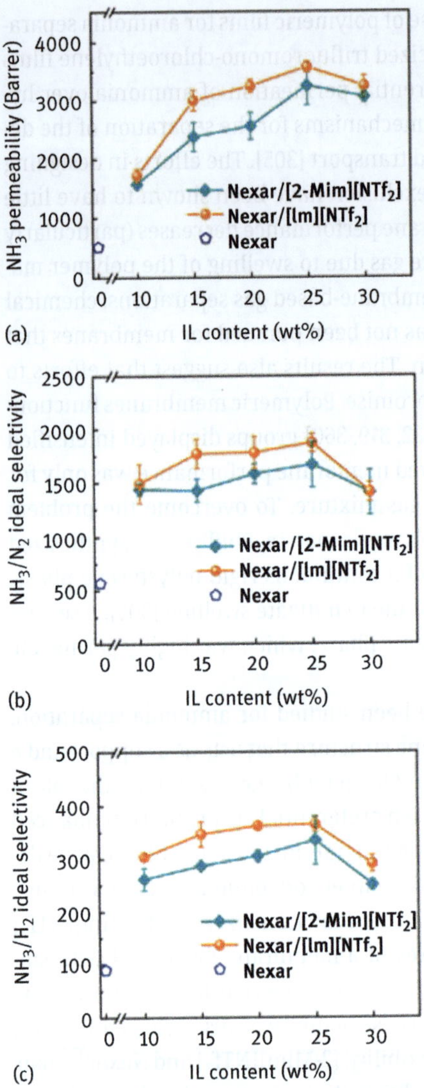

Figure 4.23: (a) Permeability of ammonia within Nexar™ and IL-impregnated Nexar™ membranes. (b) NH_3/N_2 and (c) NH_3/H_2 selectivity using Nexar™ and IL-impregnated Nexar™ membranes. Reprinted from Journal of Membrane Science, Volume 628, Bingbing Yang et al., Super selective ammonia separation through multiple-site interaction with ionic liquid-based hybrid membranes, Article 119264, Copyright 2021, with permission from Elsevier.

membranes. However, the major challenge with SLMs is the limited stability of the membrane. The pore size and liquid filler properties should be carefully optimized to increase the blowout pressure of the membrane.

4.5.2 Inorganic membranes

For effective ammonia separation at elevated temperatures exceeding 300 °C, only thermally robust and stable inorganic membranes are suitable [199]. The separation mechanisms of these membranes involve a combination of size-based separation (molecular sieving), different diffusion rates, and preferential adsorption due to charge-dipole interactions [51, 180]. Wei et al. synthesized zeolitic imidazolate framework membranes (ZIF-21) with LTA topology via seeded secondary growth, and the membranes were evaluated for the separation of ammonia from the synthesis gas [400]. The membranes showed a reasonable ammonia permeance of 1,727 GPU with selectivity of 35 and 12 for ammonia/nitrogen and ammonia/hydrogen, respectively. This separation performance was attributed to the enhanced adsorption (ammonia vs. hydrogen) and diffusivities (ammonia vs. nitrogen); ammonia interactions with the linker polar channels were identified as the key parameter influencing the separation. Duan et al. synthesized MFI membranes that showed promising permeance 3280 and 6870 GPU and separation factors of 2236 and 307 for ammonia/nitrogen and ammonia/hydrogen mixtures at ambient temperature and 300 kPa for an equimolar gas stream [84]. The major challenge with the membrane was the deterioration of the membrane under elevated temperature. Increasing the temperature to 100 °C led to approximately 100 to 200 times decrease in the separation factor with marginal improvements in membrane flux. The MFI membranes were not subjected to the actual simulated synthesis gas mixtures; thus, it is not clear whether MFI membranes would display ammonia selectivity with feed composition similar to the Haber–Bosch reactor effluent. As visualized in Figure 4.24, various ammonia-selective membranes from the literature achieve excellent performance at low temperatures. Most research is focused on studying membrane operation below 100 °C, which is not optimal for industrial conditions. Therefore, developing a high-temperature ammonia membrane (over 200 °C) is essential for improving membrane-based ammonia production. More detailed analysis and an overview of inorganic membranes can be found in the Lin et al. report [222].

Motivated by a membrane that displayed unprecedented water transport through Na^+-gated nanochannels, Padinjarekutt et al. demonstrated the performance of ceramic hollow fiber-coated NaA zeolite crystals for the separation of ammonia from synthesis gas [84, 292]. The membrane was synthesized in a two-step process; first, the fiber was seeded with a high density of nanosized NaA seeds bonded to the support. Later, a continuous, defect-free layer was synthesized by secondary growth. This is one of the few reports that evaluates the performance of the membrane with the synthesis gas and at elevated temperatures (i. e., 250 °C).

In general, membrane-based separation technologies are widely accepted as a compelling route to improve the energy efficiency of chemical processes by providing a more selective and often less energy-intensive alternative to traditional separation methods such as distillation. Furthermore, membrane systems are modular and can often be integrated directly into existing reaction pathways, enabling in situ product removal and

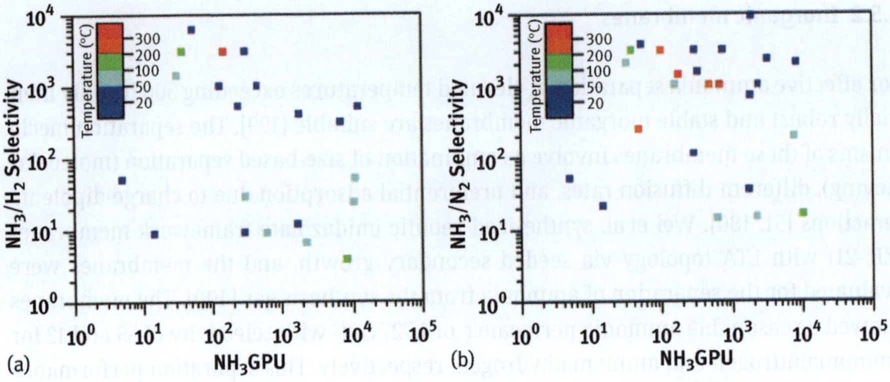

Figure 4.24: Overview of NH_3 selectivity vs. NH_3 permeance for membranes studied for (a) NH_3/H_2 and (b) NH_3/N_2 separation in literature. Reprinted with permission from Bosong Lin et al., Perspective on intensification of Haber–-Bosch to enable ammonia production under milder conditions, ACS Sustainable Chemistry & Engineering, Volume 11, Issue 27, pp. 9880–9899. Copyright 2023 American Chemical Society.

shifting equilibrium toward desired products, ultimately leading to higher yields and reduced waste, which further contributes to the overall efficiency of the chemical process. For the case of incorporation into the Haber–Bosch synthesis loop, the major concerns are: (i) optimizing the permeance and selectivity of the membranes for such challenging separation, (ii) developing robust and scalable membranes that can operate safely at elevated temperatures and pressures, and (iii) designing and evaluating a synthesis loop integrated with a membrane separation that can demonstrate the promise of membranes.

5 Cost and performance of enabling technologies for sustainable ammonia

This chapter contains a review of the state-of-the-art in terms of the cost and performance of the process technologies needed to enable renewable ammonia production. The economic and performance characteristics discussed in this chapter include capital investment, operating costs, energy efficiency, or energy intensity, and dynamic flexibility. The flexibility of processes to adjust their production in response to variability in energy availability or price can be described by three parameters: turndown, ramp rate, and startup time. Turndown is the amount by which a process can operate below its nameplate capacity and is often expressed as a percentage of this capacity. For example, a process with 80 % turndown can operate at any production level between 20 % and 100 % of nameplate capacity. Ramp rate is the speed at which a process can adjust its production rate. It is often expressed in terms of % of nameplate capacity per time. For example, if the example process described above had a ramp rate of 20 %/hour, it would take four hours to move the production rate between its fully "turned-down" 20 % lower bound and full capacity. In broad terms, the startup time is the amount of time it takes for a process to begin useful production after it has been off.

5.1 Renewable power generation

The cost of renewable generation from wind, solar, and hydroelectric resources are discussed herein. The production of ammonia from each of these resources has been explored, at least from a theoretical perspective. The capital and annual operating costs of wind turbines, PV arrays, and hydroelectric generation with projections through 2050 are given in Tables 5.1 through 5.3. These values are from the U.S. National Renewable Laboratory (NREL) Annual Technology Baseline (ATB) [269].

Wind and PV generation have a number of similarities from an economic perspective. Both exhibit economies for larger, utility-scale installations. The considerable $/MW cost savings arise from balance-of-plant and soft costs (e. g., engineering, project man-

Table 5.1: Capital and annual operating costs for renewable generation from wind turbines.

		Present day	2030	2040	2050
Distributed (<10 MW)	Capital (MM$/MW)	2.31	2.04	1.81	1.63
	Operating (MM$/MW/y)	0.039	0.038	0.037	0.036
Utility (>100 MW)	Capital (MM$/MW)	1.57	1.41	1.26	1.11
	Operating (MM$/MW/y)	0.031	0.029	0.027	0.025
Offshore	Capital (MM$/MW)	5.87	4.45	3.31	3.13
	Operating (MM$/MW/y)	0.084	0.0785	0.072	0.070

https://doi.org/10.1515/9783110762525-005

Table 5.2: Capital and annual operating costs for renewable generation from PV arrays.

		Present day	2030	2040	2050
Distributed (<1 MW)	Capital (MM$/MW)	1.73	1.44	1.05	0.86
	Operating (MM$/MW/y)	0.018	0.015	0.012	0.010
Utility (>10 MW)	Capital (MM$/MW)	1.25	1.04	0.76	0.63
	Operating (MM$/MW/y)	0.02	0.018	0.015	0.013

Table 5.3: Capital and annual operating costs for hydroelectric renewable power generation with headwater height above 10 m.

		Present day	2030	2040	2050
Power existing	Capital (MM$/MW)	3.24	3.24	3.24	3.24
	Operating (MM$/MW/y)	0.092	0.092	0.088	0.088
New build (<10 MW)	Capital (MM$/MW)	8.60	8.60	8.60	8.60
	Operating (MM$/MW/y)	0.15	0.15	0.15	0.15
New build (>10 MW)	Capital (MM$/MW)	7.60	7.60	7.60	7.60
	Operating (MM$/MW/y)	0.043	0.043	0.043	0.043

agement) rather than from the generating units themselves. This occurs because large wind and PV generation installations use repeating modules for actual generation. For example, a land-based 150 MW wind farm might consist of 100 1.5 MW turbines or a 10 MW solar project might use 50,000 0.2 kW individual panels.

Both forms of renewable generation also have uncertainty associated with their future cost reductions. In addition to the average values listed in Tables 5.1 and 5.2, the NREL ATB [269] evaluates the aggressive and conservative magnitudes of cost reduction. For distributed wind, there exists an uncertainty of roughly 25 % in either direction for the capital cost of 2050 relative to the value reported in Table 5.1. For utility-scale land-based and offshore wind generation, capital costs could be 5 % lower or 12 % higher in 2050. There is even more uncertainty associated with PV-based generation. For distributed and utility-scale PV, capital costs could be 25 % lower or 35 % higher by 2035 and 17 % lower or 33 % higher by 2050. It should be noted that the capital costs of renewable generation are subject to a number of market factors. Rapid increases in demand can result in short-term cost increase due to supply-demand dynamics before longer-term decrease due to economies of production volume. In addition, broader inflation across the economy has an effect on the costs of materials used to make wind turbines and PV arrays. For example, in 2022, the NREL ATB predicted that utility-scale wind and PV generation would have respective capital costs of $1.21MM/MW and $0.98MM/MW in 2025. Their updated prediction as of the end of 2024 displayed increases of 30 % and 22 % following global inflation during this period of time.

A third major similarity between wind and PV generation economics is the geographical variability of renewable potential and thus actual power generation costs. Since these resources are intermittent, they do not generate at full capacity throughout the year. In fact, the annual average capacity factors for all operating land-based wind, offshore wind, and solar generation in the United States have been approximately 43 %, 48 %, and 20 %, respectively [270]. This means that for 100 MW of installed capacity, the annual average generation for these generators would be 43 MW, 48 MW, and 20 MW, respectively. This is a major factor in the decision making about where to locate ammonia production powered by variable renewables. Lower renewable potential leads to higher costs for multiple reasons. Primarily, as illustrated above, more installed capacity, and thus capital investment is needed to generate the same amount of energy in locations with lower capacity factors. This leads to a higher levelized cost of energy, especially because the overall cost of renewable generation is dominated by capital investment. Additionally, a lower renewable potential generally manifests itself in more periods with little to no generation. These periods require energy storage such as batteries and/or over-sized chemical production operating in a flexible way to match renewable available, both of which add additional capital cost. Another possibility is to purchase power from an adjacent utility, although this power generally has higher associated costs than wind or solar generation, as well as higher emissions, sometimes considerably.

Hydroelectric power generation is more expensive than wind or solar generation, primarily due to high investment costs. These costs arise from the plethora of civil works such as dams and tunnels. Adding generation capabilities or "powering" existing dams can reduce this capital investment by exploiting infrastructure which has already been installed. However, studies have shown that the generation potential of "powered" existing dams is lower than that of purpose-built ones. This is because the water flow rates and/or the water height differential ("head") across the dam were not originally specified with hydroelectric generation in mind. The availability of hydropower is more consistent and stable than that of wind or PV generation. Some projects in water-rich locations can generate at >99 % capacity factors, while others have to adjust their production on seasonal timescales to account for differences in rainfall or prevent adverse water levels downstream. In the United States, the average annual capacity factors for all operating hydropower projects are about 50 % [270].

5.2 Hydrogen production

Hydrogen production is a key technological aspect of ammonia production. It certainly requires the most energy, while also contributing and sometimes dominating the capital investment required for an ammonia project. For renewable ammonia production, the two main routes are electrolysis powered by renewable energy and deriving hydrogen from biogas or biomass. Economic and performance considerations for both are discussed in this section. Special attention is also paid to hydrogen storage. Under ambient conditions, hydrogen is the least dense of all elements at 0.89 kg H_2/m^3. Storage technologies primarily serve to increase this density but can meaningfully contribute to the cost of a renewable ammonia project if high capacity is required.

5.2.1 Electrolysis

There are three main types of electrolysis: proton electrolyte membrane (PEM), alkaline (AEC) and solid oxide (SOEC) [328, 395]. Figure 5.1 provides a basic schematic of each technology. Table 5.4 provides an estimated capital cost for these electrolysis technologies from today to 2050. These projected decreases in electrolysis capital cost are largely due to economies of volume as manufacturing increases to meet the increased demand. In the short term, there is the possibility of higher electrolysis prices if demand increases faster than global manufacturing capacity. As with renewable generation, global inflation from 2022 to 2024 has led to an increase in electrolysis capital costs compared to pre-2022 projections. In general, the costs of electrolysis capital in the next 25 years are subject to considerable uncertainty. The uncertainty around the Electric Power Research Institute capital cost estimates in Table 5.4 is quantified as 40 % for alkaline electrolysis and 30 % for PEM electrolysis at present. Projecting to 2050, this uncertainty is 40 % and 25 % for alkaline and PEM electrolysis, respectively.

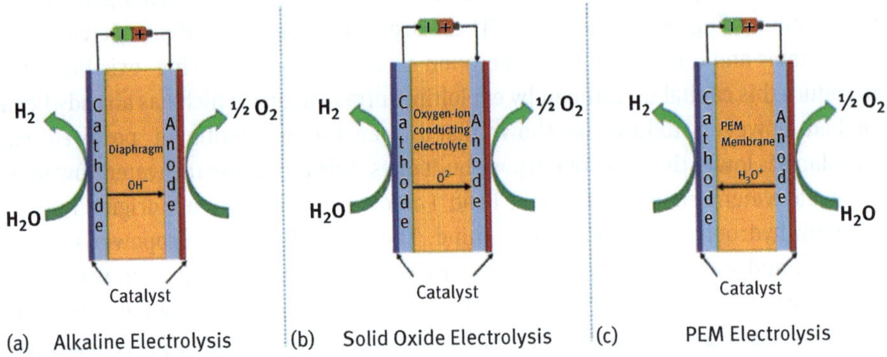

Figure 5.1: Conceptual schematic of (a) alkaline water electrolysis, (b) solid oxide electrolysis, and (c) proton exchange membrane water electrolysis. Schematic describes the respective oxidation and reduction reactions occurring at the anode and cathode as well as the anion or cation flowing between the anode and cathode for each technology. Reprinted from [395] under Creative Commons Attribution 4.0 International License.

Table 5.4: Capital costs for alkaline, proton electrolyte membrane (PEM), and solid oxide electrolysis.

Capital cost (MM$/MW)	Present day	2030	2040	2050
Alkaline (<20 MW)	1.80	1.65	1.60	1.55
Alkaline (>20 to <100 MW)	1.15	1.05	1.00	0.95
Alkaline (>100 MW)	0.80	0.70	0.65	0.60
PEM (<20 MW)	2.20	2.00	1.70	1.60
PEM (>20 to <100 MW)	1.50	1.35	1.15	1.05
PEM (>100 MW)	1.05	0.95	0.80	0.75
Solid oxide (>20 MW)	3.85	1.95	1.85	1.3

Electrolysis capital costs exhibit economies of scale despite the modular nature of increasing the number of stacks to increase the total hydrogen production capacity. These economies of scale are driven by two main factors. First, stack manufacturing can obtain economies of volume in larger electrolysis production facilities. For example, the Electric Power Research Institute (EPRI), an US nonprofit research and development organization, lists stack costs that are 50 % lower for both alkaline and PEM systems at the >100 MW scale compared to the <20 MW scale. Additionally, a considerable balance of plant (e. g., power electronics, water purification, compression) is required to actually produce usable hydrogen from electricity and water. The nonstack contributions toward the electrolysis capital cost are significant. In fact, EPRI's estimate of the balance of plant contributions is between 65 % and 75 % of the total alkaline capital costs at the <20 MW scale and 60 % to 70 % of total capital costs at the >100 MW scale. For PEM systems, the balance of plant costs represents 60 % of capital costs at the <20 MW scale and 55 % at the >100 MW scale.

Currently, alkaline electrolysis is the most widely deployed [156]. These systems operate at 50–80 °C and produce hydrogen at approximately atmospheric pressure. The system efficiency of alkaline electrolysis is approximately 55 MWh/mtH_2, or roughly 60 % based on the lower heating value (LHV) of hydrogen. This efficiency is projected to improve to 70 % of LHV by 2030 and 75 % of LHV by 2050. Alkaline electrolysis has the lowest capital cost currently due to its high technical maturity, the avoidance of expensive noble metals, and operating lifetimes of up to 90,000 hours [34]. The anode and cathode generally consist of nickel or nickel alloys [416]. The annual operating cost of alkaline electrolysis is usually 2 % to 4 % of the total capital investment. The main drawback of AEC is somewhat limited dynamic flexibility, with state-of-the-art turndown of 80 % to 90 %, ramp rates of 20 %/second, and most detrimentally, startup times on the order of 60 minutes [209].

PEM electrolysis achieves comparable 55 MWh/mtH_2 system efficiencies as alkaline systems, but offer better dynamic flexibility characteristics. Specifically, their turndown is at least 95 % and potentially up to 99 %, their ramp rates are on the order of 99 %/0.1 seconds, and their startup times are less than 20 minutes [53]. Efficiency improvements of 70–75 % of LHV are expected to be achieved by 2050. PEM electrolysis also operate between 50–80 °C but generally produce hydrogen at higher pressure than alkaline; 30 bar is the industrial standard [125]. This pressure can be as high as 200 bar, allowing synergistic coupling with gaseous hydrogen storage or typical ammonia synthesis pressures. PEM electrolysis uses rare earth metals in its cells, such as platinum (Pt) or platinum–palladium (Pt–Pd) for the anode and ruthenium oxide or iridium oxide for the cathode. This leads to a higher capital cost than alkaline systems. Furthermore, PEM electrolysis has shorter operating lifetimes, specifically in the range of 20,000 to 60,000 hours. The annual operating cost of alkaline electrolysis is usually 3 % to 5 % of the total capital investment, slightly higher than for alkaline systems. However, PEM's flexibility and resultant ability to better pair with intermittent renewables have driven industrial efforts to reduce costs and increase the lifetime. Currently, PEM electrolysis represents a moderate share of the electrolysis deployment [156], comprising approximately 30 % of the total global capacity. Many experts predict that it will be the dominant electrolysis technology deployed by 2035 or 2040 [328]. For this reason, the capital cost of PEM systems is expected to be close to that of alkaline systems between 2040 and 2050.

Solid oxide electrolysis is at a much earlier stage of development and has been mostly limited to demonstration efforts. Some MW scale projects were developed in 2022 and 2023 and commercialization is approaching [156]. These systems operate at temperatures of 650–1000 °C, pressures up to 25 bar, and can achieve efficiencies that are close to 90 % of hydrogen LHV [200]. SOEC uses ceramics such as Yttria-stabilized zirconia (YSZ) as an electrolyte, a Ni-type cathode, and perovskite-type lanthanum strontium (LSM) on the YSZ anode. The capital cost of these systems is currently high, but reductions are expected with their continued development [186]. However, high-temperature operation presents a number of challenges. Primarily, materials in solid oxide systems degrade faster than in alkaline or PEM electrolysis, leading to lifetimes less than 10,000 hours. The annual operating cost for solid oxide systems is higher as a result and generally in the range of 5 % to 7 % of the total capital investment. The dynamic flexibility of solid oxide electrolysis is somewhat limited, with turndown of 80 %, ramp rates of 80 %/second when hot, and startup times of up to an hour [130]. The key limitation to dynamic operation is that the cold start of solid oxide systems has an associated heat requirement to achieve its high operating temperature. As a result, solid oxide systems seem best suited for applications with a steady power source and opportunities for heat integration where possible.

5.2.2 Biomass-to-hydrogen

There are three main approaches to hydrogen production from biomass in the context of sustainable ammonia production. These are indirect gasification, direct gasification, and anerobic digestion. The operating conditions for each, as well as the advantages and drawbacks, are discussed in this subsection. The effluent of gasification and digestion processes is not suitable for ammonia synthesis. It must be purified and its composition further adjusted through a series of processes similar to those which occur when obtaining ammonia synthesis gas (a 3:1 molar ratio of hydrogen to nitrogen) from fossil fuels [249]. Figure 5.2 is a conceptual superstructure for the production of ammonia from biomass feedstocks.

Indirect gasification processes use two vessels with different purposes. One is used for the gasification reaction using steam to gasify biomass. The other vessel is used to burn the carbonaceous char produced during the gasification process to provide heat for the endothermic gasification reaction. A solid heat transfer medium is used to heat integrate the gasification vessel and the char combustion vessel. The gasification reaction occurs at 1.6 bar and between temperatures of 690 °C and 1015 °C. The composition

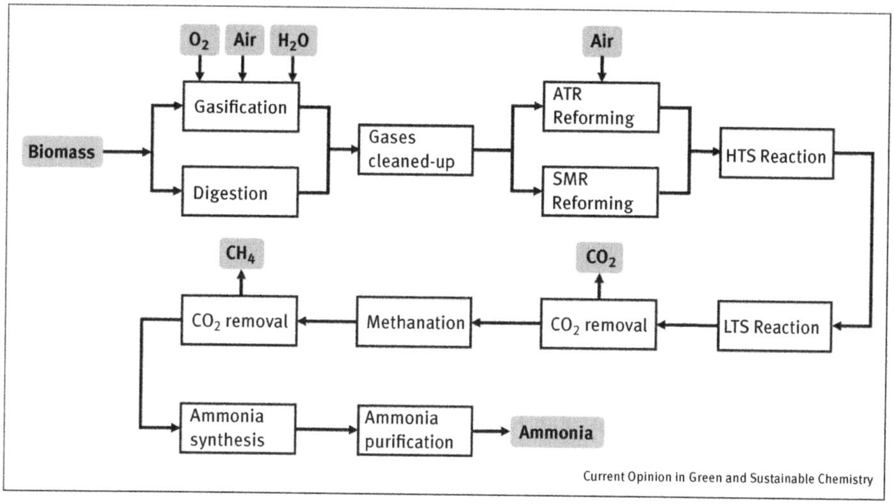

Figure 5.2: Conceptual superstructure for alternative ammonia production approaches from biomass feedstocks. The abbreviations in the figure are defined herein: ATR – Autothermal reforming. SMR – Steam methane reforming, HTS – High temperature shift; LTS – Low temperature shift. Reprinted from Current Opinion in Green and Sustainable Chemistry, Volume 47, Mariano Martin and Antonio Sanchez, Biomass pathways to produce green ammonia and urea, Article 100933. Copyright 2024, with permission from Elsevier.

of the outlet gas to be ultimately used for ammonia production (biogas) is affected by the reaction temperature. The reactor effluent consists of biogas, unreacted biomass, char, and the solid heat transfer medium. Solids are separated from biogas through a cyclone with an efficiency of 99 9 %. The capital cost of indirect gasification scales with a power-law correlation typical to the chemical industry [316, 326]:

$$\text{CAPEX}_{\text{INDIRECT}}\ (\text{MM\$}) = 22\left(\frac{\text{scale}}{500{,}000\ \text{mt}_{\text{biomass}}/\text{y}}\right)^{0.65} \tag{5.1}$$

Direct gasification uses only one vessel where steam, oxygen (whether pure or in air), and biomass are reacted to form biogas. The temperature is between 450 °C and 685 °C and again affects the composition of the biogas. As in indirect gasification, the solids that leave the gasifier are separated from the biogas through cyclones. In general, oxygen-fed direct gasification has a higher reaction pressure between 5 bar and 25 bar, a higher reaction temperature, and results in a product gas having a higher energy content. However, feeding with pure oxygen also results in additional char formation and requires a source of pure oxygen. In the context of ammonia production using biomass as the hydrogen source, this oxygen could be sourced as a byproduct of the air separation process used to produce nitrogen. Air-fed direct gasification can operate as low as atmospheric pressure and have more operational simplicity than the pure oxygen process. The biogas from the air-fed process generally contains more hydrogen and carbon monoxide but less methane and carbon dioxide, as well as a considerable amount of nitrogen, which must be balanced to the correct synthesis reaction stoichiometry in the process of making ammonia. The capital cost of direct gasification is higher than that of indirect gasification:

$$\mathrm{CAPEX_{DIRECT}}\ (\mathrm{MM\$}) = 53\left(\frac{\mathrm{scale}}{500{,}000\ \mathrm{mt_{biomass}/y}}\right)^{0.7} \tag{5.2}$$

Anerobic digestion involves the production of methane, carbon dioxide, and water through a reaction of biomass in the absence of oxygen. This process operates near atmospheric pressure and at a temperature of 55 °C. Only a portion of digested biomass is converted to biogas while the balance is converted to a solid digestate. On a mass basis, as high as 80 % of the biomass can form digestate. This digestate may have economic value as an organic fertilizer, compost material, or even as a feedstock for the production of bioplastics after further processing and treatment. However, markets for these digestate-derived products are not currently well-established. From a capital investment perspective, digestion scales different from the gasification processes described above. Scaling up digestion involves the repeated installation of normalized digester sizes. A recent study quoted a normalized digester capacity of 6,000 m^3 with a cost of approximately \$550/m^3 [81]. Each 6,000 m^3 digester would be able to produce approximately 50,000 m^3/day of biogas.

After biogas has been produced via any of the above pathways, it must be purified with a specific focus on fine particulates and hydrogen sulfide. Particulates are removed by filtration and then zinc oxide is used to remove hydrogen sulfide. Biogas produced by anerobic digestion contains less particulate matter but must be heated via an external fuel source before reforming due to its lower temperature compared to gasification effluent. Biogas must also be compressed prior to reforming, whether autothermal reforming or steam reforming is used.

Autothermal reforming (ATR) involves the adiabatic reaction of biogas, steam, and air. This reaction occurs at pressures of 20 bar to 30 bar and temperatures of 850 °C to 1000 °C. The heat required for the reforming reaction is obtained from the combustion of some of the biogas feed in the cofed air. The ratio of the amount of air included in the ATR is determined by the amount of nitrogen required to ultimately result in a stoichiometric 3:1 molar ratio of nitrogen to hydrogen fed to ammonia synthesis. ATR converts approximately 45 % of biogas to hydrogen on an energy basis [257]. The capital cost of ATR is given by the following power-law equation [75]:

$$\mathrm{CAPEX_{ATR}}\ (\mathrm{MM\$}) = 45\left(\frac{\mathrm{scale}}{350{,}000\ \mathrm{mt_{biogas}/y}}\right)^{0.67} \tag{5.3}$$

Steam reforming (SMR) involves the endothermic reaction of biogas with steam in the absence of oxygen. This reaction occurs at pressures of 20 bar to 30 bar and temperatures of 800 °C to 875 °C. A furnace-type reactor is used to supply heat to the catalytic reforming via combustion of some of the produced biogas as well as nonhydrogen components removed downstream of the steam reformer. A high steam-to-biogas ratio is generally used in an ammonia production context to maximize hydrogen production while also minimizing carbon deposition ("coking") on the reforming catalyst [257]. This comes with the drawback of a higher heat requirement to ensure isothermal conditions. In general, SMR converts approximately 60 % of biogas to hydrogen on an energy basis. From a capital cost standpoint, SMR is more expensive than ATR:

$$\mathrm{CAPEX}_{\mathrm{SMR}}\ (\mathrm{MM\$}) = 65\left(\frac{\mathrm{scale}}{350{,}000\ \mathrm{mt}_{\mathrm{biogas}}/\mathrm{y}}\right)^{0.67} \tag{5.4}$$

A water-gas shift process section is included after reforming to increase the amount of hydrogen in the reformed biogas. In both of these reactors, carbon monoxide and water are converted to carbon dioxide and hydrogen. This section generally includes two reactors in series. The first high-temperature water-gas shift reactor operates between 300 °C and 500 °C. The second low-temperature reactor operates between 180 °C and 260 °C. The water-gas shift reaction is exothermic and is equilibrium limited. The higher temperature range in the first reactor achieves a higher kinetic rate, while the lower temperature in the second reactor further shifts the equilibrium toward hydrogen. The pressure in both reactors does not affect the equilibrium conversion, but a higher pressure can serve to increase the reaction rate. The pressure in the higher temperature reactor is between 10 bar and 60 bar while the pressure in the lower temperature reactor is between 10 bar and 40 bar. The water-gas shift section effluent contains up to 70 % hydrogen on a molar basis. The capital cost of the water gas shift section is approximately [75]:

$$\mathrm{CAPEX}_{\mathrm{WGS}}\ (\mathrm{MM\$}) = 16.5\left(\frac{\mathrm{scale}}{300{,}000\ \mathrm{mt}_{\mathrm{biogas}}/\mathrm{y}}\right)^{0.67} \tag{5.5}$$

Carbon dioxide is removed via either absorption into a solvent (e. g., Rectisol or methyldiethanolamine) or adsorption in a fixed bed (e. g., activated carbon). Unremoved carbon dioxide and carbon monoxide, both of which are generally present in amounts less than 3.5 % on a molar basis, react to form methane in an adiabatic methanation reaction to lower their respective concentrations below 10 ppm. This is especially important for carbon monoxide, which will poison the ammonia synthesis catalyst. Methanation operates at pressures of 10 bar to 30 bar and temperatures between 250 °C and 350 °C. The newly formed methane, as well as any remaining from the initial biogas, is removed via adsorption. Overall, the capital cost of CO_2 removal, methanation, and methane removal is approximately [75]:

$$\mathrm{CAPEX}_{\mathrm{PUR}}\ (\mathrm{MM\$}) = 75\left(\frac{\mathrm{scale}}{300{,}000\ \mathrm{mt}_{\mathrm{biogas}}/\mathrm{y}}\right)^{0.72} \tag{5.6}$$

5.2.3 Hydrogen storage

Hydrogen storage is of crucial importance for renewable ammonia production. Key technological considerations are the cost of the storage system, the energy required to transform hydrogen into its stored form, and, specific to ammonia production, the pressure at which the hydrogen is stored. Conventional condenser-based ammonia production operates above 100 bar, so storage pressures below this would require additional compression. This section focuses on three types of hydrogen storage technologies: compressed gas, cavern, and metal hydrides. The key characteristics of each are summarized in Table 5.5.

Table 5.5: Comparison of hydrogen storage technologies.

	Compressed	Cavern	Metal hydride
Pressure (bar)	200–1,000	100–200	20 (Release[a]: 1–5)
Temperature (°C)	25	25	25 (Release[a]: 200–300)
Energy density (MWh/m^3)	0.5–1.8	0.3–0.8	1.4–1.8
Energy use for storage (MWh/mtH_2)	<4	<1	<6
Storage cost (MM$/mt)	0.5–0.8	0.035–0.05	0.025–0.035

[a]Metal hydride storage materials require decreased pressure and increased temperature to release the stored hydrogen.

Compressed gas is the most developed form of hydrogen storage. This entails ambient temperature hydrogen compressed to at least 200 bar and up to 1,000 bar, with standardized storage vessels often rated at 200 bar and 700 bar [29]. These vessels tend to be less expensive if they are located underground; this tends to be especially the case for larger inventories of hydrogen. Compressing hydrogen to 700 bar consumes approximately 10 % of the energy contained (based on LHV). The inventory of operational hydrogen storage cannot be emptied below atmospheric pressure. For a 200 bar storage, this corresponds to 0.5 % of installed capacity, while for 700 bar storage, this lower bound corresponds to approximately 0. 15 % of capacity.

Salt caverns can be used for stationary hydrogen storage on larger scales. Salt is inert with respect to hydrogen, and thus limits its diffusion from storage. Salt cavern storage generally occurs at pressures up to 200 bar, which requires less energy for hydrogen compression [290]. This storage method becomes more economical than vessel-based storage if more than 20 mtH_2 are to be stored. The specific investment is roughly an order of magnitude lower for storage volumes greater than 100 mtH_2 [13]. Salt cavern storage inventories must remain above 0. 5 % to 1 % of capacity to maintain atmospheric pressure. The main drawback of salt cavern storage is their geographic limitation in that these caverns may not exist in desirable areas for renewable ammonia production. Furthermore, hydrogen transportation to and from these caverns is nontrivial, making their use generally uneconomical for even moderate distances between the caverns and hydrogen production sites.

The storage of hydrogen in metal hydrides is based on exothermic hydrogen chemisorption onto a metal surface followed by absorption into the metal [29]. This absorbed hydrogen can be released in an endothermic process by increasing the temperature. The suitability of metals for hydride-based storage is largely dependent on thermodynamics. These dictate the temperature swing needed between absorption and release, which determines the energy penalty associated with storing hydrogen. They also affect the maximum amount of hydrogen that can be stored, which determines the total size and weight of the storage system. Magnesium hydride (MgH_2) is a promising storage candidate as it can store more than 7 % hydrogen by weight, about three times the amount stored in typical metal hydrides [264]. This storage can be achieved at pressures below 20 bar and is inexpensive from a raw material perspective, resulting in capital costs below $35,000/mt [87]. Using Mg powders with micrometer-scale particles and doping these particles with approximately 1 mol catalyst can allow charge and discharge on the order of minutes [210]. Stability and performance for this type of hydrogen storage have been demonstrated for over 2000 absorption-release cycles [237]. The main drawbacks of MgH_2 storage are the high release temperatures of 200–300 °C [320] and the low release pressure below 5 bar. Release temperatures of this magnitude require heating to use the stored hydrogen. This type of hydrogen storage could perhaps be integrated with the exothermic ammonia synthesis process in order to provide the required heat. However, the low release pressure means that significant compression would be required to use this hydrogen for ammonia synthesis. Ultimately, a whole system analysis should be performed to assess the trade-off between hydrogen storage cost and storage conditions to determine the suitability of hydride-based storage for renewable ammonia production.

5.3 Nitrogen production

Nitrogen for the production of renewable ammonia can be obtained from technologies that are already used in the air separation business. Three different types of processes are generally used: cryogenic distillation, pressure-swing adsorption, and membranes. Each technology has trade-offs with respect to scale, capital cost, energy use, and dynamic flexibility. The annual operating cost for all three tends to be 3 % to 5 % of installed capital cost.

Cryogenic distillation produces nitrogen as the distillate and oxygen as bottoms. The process operates at a temperature slightly below the boiling point of O_2 (<-183 °C) and uses at least two columns in series operating at descending pressures from 6 bar to 1.5 bar to obtain ultra-high nitrogen purity above 99.9 % [40]. The capital cost of cryogenic distillation depends heavily on production scale with the following correlation:

$$\text{CAPEX}_{\text{CRYO}}\ (\text{MM\$}) = 30\left(\frac{\text{scale}}{200{,}000\ \text{mtN}_2/\text{y}}\right)^{0.5} \tag{5.7}$$

Tight heat integration enables energy intensity of approximately 0.12 MWh/mtN_2, less than PSA or membranes, despite its operation at very low temperatures [40]. This integration also inhibits the dynamic flexibility of the process. It has a turndown of 40 %, ramp rates of 10 %/hour, and requires multiple hours for startup [254]. However, this process is the industrial standard above production scales of 100,000 mtN_2/year. The Linde Ammonia Concept (LAC) already uses cryogenic distillation to obtain nitrogen feedstock, but combines it with hydrogen obtained from steam reforming [301].

PSA is a cyclic separation process that uses at least two beds of O_2-selective carbon molecular sieve absorbent [354]. PSA operates isothermally at ambient temperature with pressure cycling between 6–8 bar for adsorption and 1 bar for desorption. PSA units are compact and inherently modular, so from a cost perspective they are well suited for small production scales. For production below 50,000 mt/y, the required capital investment scales as

$$\text{CAPEX}_{\text{PSA}}\ (\text{MM\$}) = 6\left(\frac{\text{scale}}{10{,}000\ \text{mtN}_2/\text{y}}\right)^{0.8} \tag{5.8}$$

Higher cumulative production capacity using PSA would likely involve multiple modules operating in parallel. This would cost more than cryogenic distillation. PSA also has good dynamic flexibility, with turndown of 80 %, ramp rates of 80 %/hour (full capacity), and startup times less than an hour [164]. The main drawbacks of PSA relative to cryogenic distillation are a more expensive scale-up for greater production, a lower nitrogen purity around 99.8 %, and a higher energy consumption of 0.22–0.31 MWh/mtN_2 [40].

Membrane-based air separation is best suited for small production scales below 7,500 mtN_2/year [346]. This process is less well-developed than cryogenic distillation or PSA. The investment cost of this process depends roughly linearly on the quantity of membrane required, and thus the production scale:

$$\text{CAPEX}_{\text{MEM}}\ (\text{MM\$}) = 1.1\left(\frac{\text{scale}}{1{,}000\ \text{mtN}_2/\text{y}}\right) \tag{5.9}$$

Membrane separation occurs at ambient temperature and is driven by a pressure difference between the feed and the permeate, which respectively operate at 6 to 25 bar and atmospheric pressure. This process is simpler than cryogenic distillation and thus is more flexible. However, increasing pressure is required to achieve nitrogen purity above 99.5 %, which can result in energy consumption as high as 0.65 MWh/mtN_2 [40]. Post-membrane separation is also required as a result of the presence of argon.

Nitrogen storage is less expensive than hydrogen storage but is briefly discussed here. For ammonia production, the purpose of nitrogen storage is to provide a buffer between the nitrogen production process and the ammonia synthesis loop. Under equivalent conditions, the density of nitrogen is roughly 14 times that of hydrogen. Industrially used nitrogen is generally stored either as a liquid at atmospheric pressure and –195° or as compressed gas at pressures between 200 bar and 700 bar. The capital costs for liquid nitrogen storage are approximately \$1,750/ mtN_2. This storage requires approximately 1.8×10^{-4} MWh/mtN_2 in cooling energy input to maintain the cryogenic storage temperature. Liquid nitrogen storage is usually only paired with cryogenic air separation. The nitrogen effluent of this process only requires approximately 15 °C cooling to be under storage conditions. High pressure nitrogen storage capital costs are approximately \$15,000/mtN_2 to \$50,000/mtN_2. As with hydrogen, nitrogen stored in the gas phase should remain above atmospheric pressure during operation; this corresponds to lower inventory bounds between 0.1 % and 0.5 % of installed capacity. Gaseous nitrogen storage is often paired with PSA or membrane-based air separation to avoid the energy required to liquefy the gaseous nitrogen effluent from these processes.

5.4 Ammonia production

The conventional condenser-based Haber–Bosch ammonia synthesis loop can be used for the production of renewable ammonia. The capital cost of this process is affected by economies of scale. Through a review of the relevant literature, two cost-scaling correlations have been identified depending on the scale range of interest. For production scales below 100,000 mt/y [28, 261, 27, 294]:

$$\mathrm{CAPEX}_{\mathrm{HB}}^{\mathrm{small}}\ (\mathrm{MM\$}) = 4\left(\frac{\mathrm{scale}}{1{,}000\ \mathrm{mtNH_3/y}}\right)^{0.6} \tag{5.10}$$

For production scales above 100,000 mt/y [74, 97, 322]:

$$\mathrm{CAPEX}_{\mathrm{HB}}^{\mathrm{large}}\ (\mathrm{MM\$}) = 95\left(\frac{\mathrm{scale}}{250{,}000\ \mathrm{mtNH_3/y}}\right)^{0.65} \tag{5.11}$$

Annual O&M costs tend to be between 3 % and 5 % of the total capital investment. For electrically powered implementations of the condenser-based process, which are not energy integrated with a reforming-based hydrogen production process, the energy intensity is approximately 0.4 $MWh/mtNH_3$ to 0.75 $MWh/mtNH_3$ [294, 74].

The dynamic flexibility of the ammonia synthesis loop has been the subject of intense research and development over the past 5 years. This can be crucial to the viability of renewable ammonia production, not from the standpoint of energy consumed directly by the ammonia synthesis process, but rather to minimize hydrogen consumption, and thus required hydrogen buffer storage. Conventional wisdom dictates that the ammonia synthesis loop should operate in a rigid way at or near its nameplate capacity as much as possible. However, the potential importance of flexible ammonia synthesis has led to established ammonia technology licensors developing process designs and accompanying control approaches that enable turndown up to 90 % and bidirectional ramp rates of 90 %/hour [108, 361].

A major goal of innovation in the ammonia synthesis process is to reduce capital investment, especially at the small scale. An approach with recent momentum has been to replace condensation with a zeolite adsorbent [33, 234] or metal halide absorbent [247, 385] to allow a more complete separation of ammonia from unreacted hydrogen and nitrogen at lower pressure (<50 bar) and higher separation temperature (ambient or above). A Colorado-based company, Starfire Energy, has commercialized this adsorbent-based process and showed that it has up to 99 % turndown capability [33]. They plan to sell small synthesis modules in the 35 mt/y to 350 mt/y scale range [114], but cost and energy intensity data are not available at present. The University of Minnesota has championed the development of an absorbent-based process and found that capital costs can be lower than traditional condenser-based at production scales below 100,000 mt/y [294]. The main barriers to commercialization are optimizing the cyclic operation of ammonia absorption and desorption to minimize energy consumption and better understanding the dynamic capabilities of the process. Recent research efforts on process intensification have led to concepts such as membrane reactors [422] or a single-vector catalyst-sorbent system [348]. These concepts are attractive in that they reduce the required number of process units and associated capital cost, especially on a small scale. However, intensification also leads to a loss in design degrees of freedom, for example, no longer having different reaction and separation temperatures, so additional development is needed to achieve adequate production rates and energy intensities.

5.5 Energy storage and generation

Renewable ammonia production that relies entirely on variable wind and solar resources may need onsite energy storage to assist in meeting minimum power demands for nitrogen and ammonia production as well as auxiliary facility demands. Batteries can be used for direct storage of electrical energy. Previously produced hydrogen or even product ammonia could also be used for power generation. The cost and performance of batteries along with hydrogen- and ammonia-to-power technologies are described in this section. The reader should note that this discussion is also pertinent to the use of ammonia in energy applications.

The capital and annual operating costs of batteries today and projected to 2050 are given in Table 5.6. The cost of battery energy storage depends on two specifications of a given system: the energy capacity (in MWh) and the charge/discharge capacity (in MW). For instance, a battery with 100 MWh energy capacity that can only charge/discharge at a rate of 10 MW will be less expensive than one that can charge/discharge at a rate of 50 MW. The capital and operating cost of batteries should decrease considerably by 2050, specifically of the energy capacity component. There is undoubtedly some uncertainty with this cost deflation. By 2050, battery capital and operating costs could be 25 % lower

Table 5.6: Capital and operating costs for battery energy storage.

	Present day	2030	2040	2050
Energy capital cost (MM$/MWh)	0.28	0.23	0.19	0.15
Power capital cost (MM$/MW)	0.34	0.29	0.27	0.25
Energy operating cost (MM$/MWh/y)	0.0072	0.0057	0.0047	0.0036
Power operating cost (MM$/MW/y)	0.0072	0.0072	0.0068	0.0062

and 60 % higher than the values reported in Table 5.6 [269]. The charge and discharge efficiency of state-of-the-art batteries is between 90 % and 95 %, giving round-trip efficiencies in the range 80 % to 90 %. Battery inventory or state-of-charge should remain above 20 % to mitigate performance degradation over the lifetime of the battery, while the "ramp rate" depends on the chosen capacity of the charge/discharge component.

Hydrogen or ammonia can be used for power generation via either fuel cell or as fuel for combustion. Ammonia can also be decomposed into hydrogen before being used for power generation. This can offer some benefits for either power generation technology. PEM fuel cells are not directly compatible with ammonia and the combustion properties of hydrogen and ammonia are different. However, ammonia decomposition is an endothermic process that consumes energy, so this must also be factored into the selection of power generation technology for a specific application. Table 5.7 gives the capital costs for PEM and solid oxide fuel cells. Solid oxide fuel cells are compatible with either hydrogen or ammonia fuel.

Table 5.7: Capital costs for PEM and solid oxide fuel cells.

	Present day	2030	2040	2050
PEM (MM$/MW)	1.7	1.25	0.95	0.75
Solid oxide (MM$/MW)	3	2	1.4	1

The annual operating cost of PEM fuel cells can be approximated as 1 % to 3 % of the total capital investment. The efficiency of these systems is approximately 20 MWh/mtH_2 or 60 % LHV. As with PEM electrolysis, these fuel cells have excellent dynamic response characteristics. They have 99 % turndown, 99 %/hour ramping, and startup times less than 20 minutes. The decomposition of ammonia for use in PEM fuel cells requires additional process technology. The capital cost of ammonia decomposition scales as follows [299]:

$$\mathrm{CAPEX_{DCMP}\ (MM\$)} = 3.2\left(\frac{\mathrm{scale}}{1{,}000\ \mathrm{mtH_2/y}}\right)^{0.73} \tag{5.12}$$

The annual operating cost of ammonia decomposition is between 4 % and 6 % of total capital investment. The efficiency of the decomposition is up to 70 % based on LHV [244], so the overall ammonia to power efficiency using a PEM fuel cell would be roughly 40 % or 2.1 MWh/mtNH_3. The endothermic ammonia decomposition reaction occurs at temperatures as high as 800 °C. As a result, this process requires at least an hour for startup [90]. At this time, no studies have been published on turndown and ramp rates for the ammonia decomposition process.

The annual operating cost of solid oxide fuel cells can be approximated by 3 % to 5 % of the total capital investment. The electric efficiencies of these systems can be between 60 % and 70 %. For hydrogen, this corresponds to specific energy production as high as 23.5 MWh/mtH_2, whereas for ammonia, the specific energy production can be as high as 3.6 MWh/mtNH_3. In other words, the use of ammonia directly in solid oxide fuel cells causes minimal to no efficiency loss. These systems operate at temperatures of to 600 °C to 1000 °C. This allows them to achieve combined heat and power efficiencies of above 90 % when coupled with a use for this heat. However, high-temperature operation also limits flexibility. Solid oxide fuel cells have turndown up to 90 %, ramp rates of 90 %/second, but cold startup requires multiple hours before power generation can occur.

Capital and operating costs for simple cycle and combined cycle power generation systems that use natural gas as fuel as given in Table 5.8. Cost projections for these natural gas systems are well developed [269]. Using hydrogen or ammonia in combustion systems requires some design changes that increase capital investment. The exact magnitude of the increase in capital costs is difficult to predict, but in the range of 10 % to 20 %. Using either fuel, state-of-the-art simple or combined cycle turbine generation has turndown of 80 %, ramp rates of 10 %/minute to 50 %/minute, and startup times of less than 30 minutes.

Table 5.8: Lower bound capital and operating costs for simple cycle and combined cycle combustion-based power generation systems. Hydrogen- and ammonia-fueled turbine costs will likely cost between 10 % and 20 % more for the same generation capacity.

		Present day	2030	2040	2050
Simple cycle	Capital (MM$/MW)	1.32	1.27	1.17	1.07
	Operating (MM$/MW/y)	0.026	0.025	0.023	0.022
Combined cycle	Capital (MM$/MW)	1.50	1.43	1.29	1.21
	Operating (MM$/MW/y)	0.034	0.032	0.029	0.027

Hydrogen combustion has a higher flame speed than natural gas combustion, and the resulting higher temperature can lead to higher rates of nitrogen oxide formation [149]. This can be mitigated by advanced hydrogen mixing devices [52] or sequential combustion chambers with intermediate temperature control [43]. Specific material choices can also help avoid corrosion and degradation caused by hydrogen and the increased amount of steam generated by its combustion. Simple cycle turbines using pure hydrogen as fuel have demonstrated electrical efficiencies of 35 % or 11.5 MWh/mtH_2 [52, 110]. Hydrogen-fueled combined cycle turbines can achieve electrical efficiencies as high as 60 % or 20 MWh/mtH_2 by using waste heat from the direct combustion turbine to generate steam that drives a secondary turbine.

Ammonia, on the other hand, has a lower flame speed and higher autoignition temperature than conventional fossil fuels. Comixing with a combustion promoter can alleviate this [103]. One approach is to blend ammonia with 5–10 % hydrogen. This strategy can maximize efficiency without CO_2 emissions that would come from a fossil-based promoter [260]. Pure ammonia mixtures can also be preheated by exploiting the exothermic combustion to reduce or eliminate the need for blending. This strategy can only be pursued during continuous operation once the heat from combustion is available [57]; a different combustion promoter would still be needed for start-up. Another operational challenge is that ammonia combustion dramatically increases nitrogen oxide emissions. These can be removed via post-combustion selective catalytic reduction (SCR) of NOx with ammonia, which react to form nitrogen and water. The SCR unit adds cost to the combustion system. Further, the additional nonfuel use of ammonia decreases the overall or "black box" ammonia-to-power efficiency [41]. Despite these challenges, combustion turbines that use ammonia as fuel have been demonstrated [367, 194]. Ammonia combustion can achieve simple cycle efficiencies of up to 35 % LHV or 1.8 MWh/mt. The use of ammonia in combined cycle systems has been less explored at present, but recent modeling research has shown that efficiencies for pure ammonia combustion above 60 % LHV or 3.1 $MWh/mtNH_3$ are achievable [183].

5.6 Levelized cost

The preceding sections of this chapter described capital and operating costs, as well as energy efficiency and operational flexibility of different constituent technologies for renewable ammonia production. Production cost is an important metric to holistically assess the economics of renewable ammonia projects. This will capture the total capital investment, annual O&M costs, and energy costs that arise from the production of renewable ammonia while each component technology operates within its performance specifications. A commonly used production cost metric is the levelized cost of ammonia (LCOA):

$$\text{LCOA } (\$/\text{mt}) = \frac{\text{Annualized Capital } (\$/y) + \text{Operating } (\$/y)}{\text{Production Scale } (\text{mt}/y)} \tag{5.13}$$

When ammonia is used in energy applications, it can be helpful to compare costs on an energy basis. A commonly used metric for this purpose is levelized cost of energy (LCOE):

$$\text{LCOE (\$/MWh)} = \frac{\text{Annualized Capital (\$/y)} + \text{Operating (\$/y)}}{\text{Energy Demand Scale (MWh/y)}} \tag{5.14}$$

The magnitude of annualized capital investment is a crucial aspect for levelized cost calculations. This is especially true for renewable projects that are dominated by capital costs and then have little to no feedstock costs throughout their operational life. This calculation is usually achieved using a scaled project lifetime (θ) or its inverse, a capital recovery factor (CRF), which is calculated using a project lifetime (N) and weighted average cost of capital (WACC):

$$\text{Annualized Capital (\$/y)} = \frac{\text{Capital (\$)}}{\theta\ \text{(y)}} = \text{CRF}\ (\text{y}^{-1}) \times \text{Capital (\$)} \tag{5.15}$$

$$\theta = \frac{(1 + \text{WACC})^{N} - 1}{\text{WACC}(1 + \text{WACC})^{N}} \tag{5.16}$$

Project lifetimes in the chemical and energy industry generally range from 20 to 30 years. Utility-scale energy systems often have a 30-year life expectancy [373]. The WACC captures a number of project financing considerations, including the ratio of equity-to-debt financing, the cost of equity, the interest rate and loan terms for debt, and the prevailing tax rate. WACC generally varies between projects, companies, and entire industries. It can also be heavily influenced by the country or continent in which such an investment is occurring. Countries with less of a track record of project development will often have higher WACC because potential investors assume the existence of higher degree of risk. Table 5.9 gives the average WACC and the corresponding scaled project lifetime for 25-year projects in the chemical, agriculture, utility and renewable energy industries [69].

Table 5.9: WACC and 25-year scaled project lifetime for various industries.

	WACC (%)	Scaled Project Lifetime (y)
Chemical	8.95	9.86
Agriculture	9.07	9.77
Utility	5.9	13.97
Renewable Energy	8.93	9.88

6 Sustainable ammonia production economics

Renewable ammonia can be produced from a variety of feedstocks as outlined in Chapter 2. This chapter discusses the present-day economics of renewable ammonia production from (i) variable wind and solar generation, (ii) hydropower, and (iii) biomass. It also includes a case study on renewable ammonia production in southwest Minnesota using wind and solar resources. For any of these feedstocks, the individual technologies required to produce ammonia (power generation, electrolysis, and ammonia synthesis) are already commercially available. However, they largely have not been deployed yet together for the purpose of making renewable ammonia. Furthermore, some technologies are experiencing a rapid increase or expected increase in demand. The cost of these technologies is thus subject to uncertainty and, therefore, so is the cost of making renewable ammonia.

6.1 Ammonia production from wind and solar

Ammonia production facilities that use wind and solar generation (see Figure 6.1) must be designed and/or operated to accommodate the variable and intermittent nature of these resources. Theoretically, there are multiple approaches to achieve this, including (i) using batteries or other energy storage to allow all chemical production to operate at steady state, (ii) operating electrolysis in a generation-following way and using hydrogen buffer storage to enable steady-state ammonia production, or (iii) operating all chemical production in a flexible manner to minimize onsite storage of energy or hydrogen, and thus oversizing it to ensure annual production targets are met. In practice, each of these sole approaches has drawbacks, from the high cost of battery or hydrogen storage to

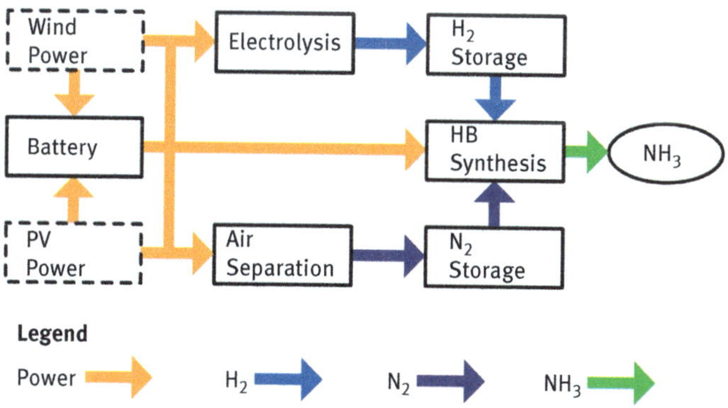

Figure 6.1: Conceptual flowsheet for renewable ammonia production from wind and solar resources. Power generated from wind and solar resources is used to separate hydrogen from water via electrolysis, nitrogen from air, and to combine these feedstocks to make ammonia. Electrical energy can be stored in a battery while hydrogen and nitrogen can be stored before being used to produce ammonia.

https://doi.org/10.1515/9783110762525-006

the technical limitations of nitrogen or ammonia production to being start-up and shut-down frequently.

A systematic approach to economically optimal design of renewable ammonia production facilities is through combined design and scheduling optimization models. These models minimize the cost of ammonia production, or more generally minimize or maximize another economic objective, by simultaneously optimizing (i) the selection and sizing of renewable generation, batteries and chemical production technologies, and (ii) the hourly operation of batteries, chemical production, and chemical storage. Including scheduling optimization at the design stage allows the necessary flexible operation considerations to be explicitly accounted for in the degree to which different technologies are oversized and the capacity of each type of storage to be installed. For more technical details about combined design and scheduling optimization formulations, readers are directed to a review paper recently written by the authors Palys and Daoutidis [296].

Two recent studies explored the economics of continuous ammonia production from solar generation, using flexible electrolysis as well as battery and hydrogen storage to achieve this. Osman et al. [287] determined the LCOA for a 600,000 mt/y continuous ammonia plant in the UAE powered by PV power where the annual average solar capacity factor is greater than 30 %. They optimized the selection, sizing, and scheduling of upstream electrical and thermal energy storage along with hydrogen and nitrogen production and storage infrastructure to enable this constant operation. They assumed battery capital costs of \$220/kWh, hydrogen storage cost of \$500/kg, and alkaline electrolysis with capital cost between \$650–710/kW. They found present-day LCOA below \$720/mt, with LCOA reduction to \$450/mt by 2030. Bouaboula et al. [44] also determined the optimal LCOA for close to continuous ammonia production in four Moroccan locations with annual average PV capacity factor greater than 30 %. They analyzed a 1,500 mt/y plant. They assumed present-day battery capital costs of \$250/kWh for storage capacity (plus \$132/kW for charge/discharge) and PEM electrolysis capital cost of \$770/kW. These assumptions resulted in LCOA below \$775/mt. Using \$34/kWh (and \$34/kW) battery costs and \$380/kW PEM electrolysis costs projected for 2050, they determined that LCOA could be reduced to \$250/mt. In general, the viability of continuous ammonia production from renewable resources appears restricted to locations with good PV potential. Both UAE and Morocco are close to the equator, meaning that their seasonal variation in solar potential is minimal. This is important because long-duration battery or hydrogen storage would make this continuous production concept noncompetitive.

Connecting ammonia production to existing grid infrastructure can help reduce production costs while potentially limiting the amount of dynamic operation required. Armijo and Philibert [22] determined optimal LCOA for 35,000 mt/y ammonia production in four South American locations with high wind (CF>40 %) and solar potential (CF>29 %). Their system did allow for the purchase of "firming" power from the local utility. It used alkaline electrolysis with capital cost of \$600/kW and hydrogen storage with capital cost of \$400/kg. The assumed ammonia synthesis had turndown and ramp parameters of 40 % and 20 %/h, respectively. They found LCOA below \$750/mt in all locations. These systems used between 4 % and 15 % grid power to mitigate hydrogen storage costs. The lowest LCOA below \$500/mt is achieved in Taltal, Chile, which has annual average wind and solar capacity factors of 43.8 % and 32.5 %, respectively. They also determined that if the ammonia synthesis turndown could be improved to 80 %, LCOA could be reduced to below \$575/mt in all locations. Salmon and Banares-Alcantara [321] analyzed 1,000,000 mt/y renewable ammonia production in Australia, coupled with options to buy and sell power with the grid at time-varying prices. They considered capital costs of \$700/kW and \$500/kg for electrolysis and hydrogen storage, respectively. They conducted case studies for six Australian states and found LCOA between \$700/mt and \$900/mt, with this consumer-supplier grid interaction giving 5 % a average LCOA reduction. In addition, using ammonia production as a demand response consumer and selling power to the grid also reduced power supply emissions in four of the six states. Campion et al. [50] determined LCOA for 430,000 mt/y production in Northern Chile, Denmark, and South Australia. They considered both islanded and grid connected systems and determined production costs for systems, which used alkaline (AEC) and solid oxide electrolysis (SOEC) with capital cost of \$925/kW and \$3,000/kW, respectively. They assumed hydrogen storage costs of \$500/kg. For islanded wind/PV systems, they found LCOA between \$900–1,200/mt using AEC. Production costs were at least 35 % using SOEC mainly due to the higher electrolysis CAPEX. Grid-connected systems reduced ammonia production costs by up to 25 % by allowing higher electrolysis capacity factors and thus reducing capital expenditure.

Bose et al. [42] determined LCOA for grid-connected and off-grid ammonia production throughout the United States under the constraint that ammonia production must be operated continuously. They used 2030 projections for capital cost, notably \$500/kW for PEM electrolysis and \$345/kg for hydrogen storage. The costs of continuous ammonia production when connected to the grid were as low as \$540/mt, while the costs of continuous ammonia production outside the grid using dedicated renewable generation plus storage were at least \$1,000/mt. However, they also found that ammonia produced in grid-connected systems can have a widely ranging emissions intensity depending on the emissions intensity of the grid itself. For example, using grid-powered ammonia production in Amarillo, Texas, would have 80 % lower emissions intensity than state-of-the-art (SOTA) methane-based production, while using the grid in Greenfield, Indiana, would actually result in higher emissions intensity than current SOTA production. It is thus imperative to carefully track the emissions intensity of the electricity being used in grid-connected ammonia projects and for this to be explicitly considered into how "green" such ammonia truly is.

Combining wind and PV resources can help improve the economics of 100 % renewable or "islanded" ammonia production. Nayak-Luke et al. [272] determined renewable ammonia production costs for a 85,000 mt/y production scale in Lewrick, Scotland.

They assumed that their facility could purchase wind and solar energy from existing generation and used 5 years of historical hourly resolution renewable availability data to optimize the design and operation of their considered facility. They considered alkaline electrolysis with a capital cost of $1,100/kW, and hydrogen storage with a capital cost of $480/kg. They conducted case studies to determine the effect on the optimized ammonia production costs of the wind/solar generation fractions, renewable electricity costs, and the ammonia synthesis turndown and ramp rate parameters. They found that a 9:1 wind-to-solar ratio and 80 % and 6 %/hour synthesis turndown and ramping gave an optimal LCOA of $715/mt. Wang et al. [387] determined optimal ammonia production costs using 2030 projections for technology costs in 21 Australian locations for wind, PV, and hybrid wind-PV energy sourcing. Their ammonia production facility considered alkaline electrolysis with a capital cost of $485/kW and turndown of 80 %, $600/kg$H_2$ hydrogen storage capital cost, and 40 % turndown for Haber–Bosch synthesis. They found an ammonia production cost for 2030 as low as $660/mt in the location that had the best combined wind and PV potential. Florez et al. [99] investigated the cost of installing new wind and solar power generation to produce ammonia in three coastal locations in Saudi Arabia. Their ammonia production facility included PEM electroysis with a capital cost of $341/kW and turndown of 100 %, $500/kg$H_2$ hydrogen storage capital cost, and 80 % turndown for ammonia synthesis. Their facility considered the possibility of both battery storage and hydrogen fuel cells to meet minimum power demands for nitrogen and ammonia production. They found a minimum ammonia production cost in 2030 of $400/mt on the northwest coast of Saudi Arabia. Laimon and Goh [201] performed a similar analysis for Jordan, but specifically focused on the use of otherwise unused ("curtailed") renewable energy. They found that Jordan had curtailed an average of 23 % and 27 % of wind and solar energy, respectively, from 2018 to 2022, and this curtailment rate is projected to increase as renewables further proliferate their power sector. Their ammonia production analysis considered the use of PEM electrolysis with a capital cost of $1,060/kW and found ammonia production costs in the range of $400/mt to $600/mt.

Three efforts have been made to develop global LCOA atlases for ammonia production from 100 % wind and PV resources. Nayak-Luke et al. [274] extended their previous analysis from [272] to consider 534 locations around the world. They assumed the use of alkaline electrolysis with a current capital cost of $700/kW and projections to $344/kW by 2030. The cost of hydrogen storage was taken to be $500/kg. Under these assumptions, it was determined that renewable ammonia could be produced for less than $750/mt in the best locations today, with projections of less than $400/mt by 2030 at an approximate production scale of 85,000 mt/y or 100 MW annual average power supply. Fasihi et al. [97] created an optimal ammonia production cost atlas on a 0.45° by 0.45° spatial grid. They assumed a capital cost for alkaline electrolysis of $750/kW today with a reduction to $275/kW by 2050 and hydrogen storage costs of $300/kg in underground pipelines and $10/kg in salt caverns in certain locations where available. They also considered fully flexible operation (i. e., 100 % turndown) for all processes except ammonia synthesis, which had 50 % turndown and hourly ramp up and down limits of 2 % and 20 %, respectively. They considered a 400,000 mt/y ammonia production scale and used technology cost projections from today to 2050. They found LCOA below $700/mt today in the best locations worldwide, with a projection below $300/mt by 2050. Fernandez et al. [98] added a dimension to the above type of global analysis by also considering water stress. This analysis used a $500/kW capital cost for PEM electrolysis. They determine an average ammonia production cost of $700/mt across a selection of the 144 most economical locations when optimizing only for an economic objective. Optimizing to minimize water stress increases the average production cost to $820/mt when considering the 144 locations with the lowest water stress. This analysis showed a broad trade-off between cost and water stress because the locations with the best renewable production may not be the locations with suitable water availability. However, it also found over 70 % locations common to both minimization approaches, indicating that low-cost production can still be found while also accounting for water stress.

Two recent research efforts have also investigated the possibility of using offshore wind generation for ammonia production. Wang et al. [389] performed the first of these studies. They assumed $1,300/kW and $110/kg capital costs for electrolysis and hydrogen storage, respectively. In addition to typically considered batteries and gas buffer storage, they allowed the use of hydrogen fuel cells and ammonia ICE generators to provide backup power. They compared the economics of onshore ammonia production connected to offshore wind by underwater power cables with offshore ammonia production, and found offshore production to be less expensive for 1,500 km distances from land. Optimal offshore ammonia production designs in fact used hydrogen fuel cells to act as a dispatchable power source. The LCOA for 500,000 mt/y ammonia production was found to be roughly $1,000/mt for average wind speeds of 11 m/s and water depth less than 100 m, with projected improvement to less than $700/mt by 2050. Much of the capital investment for this offshore production is for the required platforms for wind turbines and chemical production processes. Salmon and Banares-Alcantara also analyzed the economics of ammonia production from offshore wind [322]. They assumed a capital cost for electrolysis and hydrogen storage of $825/kW and $500/kg, respectively. They determined the LCOA of both onshore and offshore ammonia production with a scale of 1,000,000 mt/y. The best-case offshore ammonia production costs were approximately $750/mt. In general, they found that offshore production costs are higher than land-based wind production costs by up to 40 %, but that global land use restrictions may still create a favorable economic climate for offshore production.

A major development in economic assessments for renewable ammonia production beginning in 2024 is more detailed investigation into how the flexibility characteristics of ammonia synthesis systems affect the economic and emissions intensity (in grid-connected systems) of renewable ammonia production. Wu et al. [406] conducted a detailed study to determine the effect of the ammonia synthesis loop (i) turndown, (ii) ramp rate and (iii) adjustment interval, which they defined as the time required at a given production rate before changing to a new one. Their analysis considered the use of wind and solar resources in 100 % renewable and grid-connected facilities located in Xinjiang, China. Their facilities used alkaline electrolysis with a capital cost of \$205/kW and hydrogen storage with a capital cost of \$500/kg. For grid connected scenarios, they assumed that power could be purchased at a fluctuating price based on a historical average, while excess wind or solar energy could be sold for \$36/MWh. This work confirmed that grid-connected ammonia production is considerably cheaper than 100 % renewable islanded production for present-day alkaline electrolysis (especially compared to PEM systems) due to the high amount of energy storage needed to meet minimum power demands for energy-intensive electrolysis processes. This work also found that for grid-connected systems the following flexibility characteristics of the synthesis process are suitable in terms of both the production cost and the hydrogen storage capacity in that more aggressive targets give diminishing returns: 70 % turndown, a ramp rate of 30 %/hour, and a daily (24 hour) adjustment frequency. This study confirms that the synthesis flexibility claims made by multiple technology licensors are suitable for on-grid applications. However, replicating such a systematic study for off-grid ammonia production using PEM electrolysis would be valuable.

Mingolla et al. [255] performed a systematic study of the effect of the flexibility of the synthesis in relation to the wind and solar potential in ammonia synthesis facilities connected to the grid. They modeled the production cost at facilities with continuous and flexible synthesis loops (with 50 % turndown) for five representative synthetic regions with different wind and solar potential. They consider the use of alkaline electrolysis with \$1,200/kW capital cost, \$500/kg hydrogen storage, battery storage for onsite backup power and grid electricity with an average price of \$115/MWh, an emissions intensity of 0.15 $mtCO_2$/MWh, and an emissions intensity limit arising from grid purchases of 0.175 $mtCO_2/mtNH_3$. Ultimately, this study found that continuously operating ammonia synthesis facilities offer lower production costs in regions dominated by solar energy, while synthesis flexibility is more valuable in regions with high wind potential. The diurnal characteristics of solar generation, especially in regions near the equator, lead to an optimal operational strategy of producing surplus hydrogen during the day to allow continuous operation overnight. However, more random and seasonal fluctuations in wind dominant regions mean that flexible synthesis is imperative to reduce hydrogen storage costs. In general, this work reflects the results of the majority of recent technoeconomic studies wherein the inclusion of wind energy leads to the lowest renewable ammonia production costs.

Zhou et al. [430] analyzed the effect of detailed flexibility in ammonia synthesis on the ammonia production cost. They compared the costs of ammonia production in Inner Mongolia, China arising from facilities with steady-state synthesis, continuously flexible synthesis with an operating range 30 % to 100 %, and synthesis with multiple discrete stable operating points from 30 % to 100 % in increments of 10 %. Their studied facility considered the use of both wind and PV generation resources, alkaline electrolysis with \$285/kW capital cost and 80 % turndown, \$550/kg hydrogen storage capital cost, and the ability to both buy power from and sell power to an adjacent utility with respective prices of \$53/MWh and \$41/MWh. They assumed that the purchased power supply had a carbon itensity of 0.5703 $kgCO_2$/kWh. They found that both flexible operation approaches resulted in ammonia production costs below \$500/mt at production scales slightly above 70,000 mt/y. More specifically, a continuously flexible synthesis loop results in a slightly lower ammonia production cost and minimal carbon intensity (0.01 $kgCO_2/kgNH_3$) by minimizing the amount of power purchased from the grid. However, multimode flexibility resulted in five times less volatility in daily ammonia production rates and half the amount of electrolysis cycling on an annual basis in a trade-off of only 2.5 % higher ammonia production cost and emission intensity still below 0.1 $kgCO_2/kgNH_3$. This study raises the interesting question of how to balance cost and operational volatility for both grid-connected and 100 % islanded renewable ammonia production.

Yu et al. [413] analyzed renewable ammonia production in Inner Mongolia, China and explicitly considered 20 years of historical wind and solar generation data in their economic modeling to account for the interannual variability of these resources. Their analysis considered alkaline electrolysis with a capital cost of \$500/kW, a \$400/kg hydrogen storage capital cost, and both batteries and hydrogen fuel cells as onsite dispatchable power sources. They found that an ammonia synthesis loop, which can adjust its production rate daily, would enable an average annual production cost of \$500/mt at an annual average production scale of 100,000 mt/y. The inclusion of characteristics of multiple years of data led to the lowest average renewable ammonia production cost by simulating the optimally designed facility across each of the 20 years of historical data used in that study.

6.1.1 Case study: Renewable ammonia production in Minnesota

This subsection contains a case study on the economics of producing renewable ammonia in Minnesota using wind and solar resources. The LCOA for a number of different renewable ammonia production scenarios is minimized through the combined design and scheduling optimization approach described at the beginning of the chapter. The model uses hourly time-series data for the wind and PV capacity factors [271] as input data.

Figure 6.2 compares the LCOA of rigid vs. flexible ammonia production for a 100,000 mt/y wind-to-ammonia facility in Southwest Minnesota. Table 6.1 compares the optimal installed capacity of each constituent technology. The annual average wind capacity factor at this location is 48 %. Flexible ammonia production is clearly an enabling technology for economical renewable ammonia. The flexible system achieves 30 % lower LCOA compared to a rigid system in which ammonia must be produced at a constant rate. This cost reduction is achieved primarily through lower investment in hydrogen storage, as well as PV and wind generation. A flexible synthesis system reduces the required onsite hydrogen storage capacity by 94 %. The rigid synthesis system

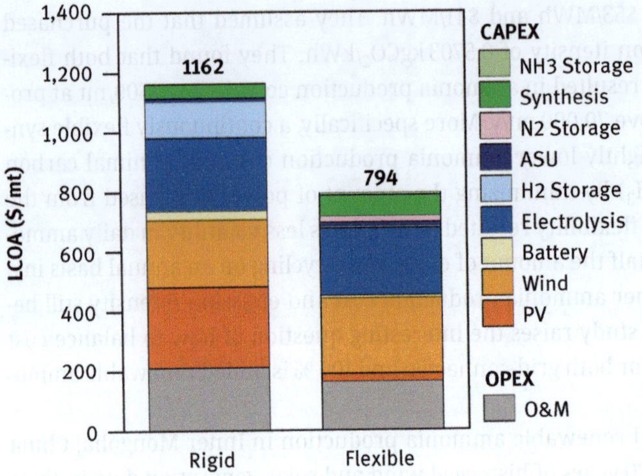

Figure 6.2: Levelized cost of renewable ammonia for 100,000 mt/y production scale in Southwest Minnesota using rigid and flexible ammonia synthesis loop technologies. The rigid ammonia synthesis loop must produced ammonia at a constant rate while the flexible ammonia synthesis loop can turndown its production to 10 % of its installed capacity. All technology cost and performance data used to generate these results are listed in Chapter 5. O&M costs comprise annual maintenance costs as well as component replacement where required such as electrolysis stacks or ammonia synthesis catalyst.

Table 6.1: Optimal installed capacities for renewable ammonia production facilities with rigid and flexible ammonia synthesis loop technologies.

Technology	Rigid	Flexible
Wind (MW)	215	241
PV (MW)	315	34
Battery (MWh)	125	41
PEM electrolysis (MW)	243	224
H_2 storage (mtH_2)	180	11
Air separation (ktN_2/y)	83	97
N_2 storage (mtN_2)	49	2,510
NH_3 synthesis ($ktNH_3/y$)	100	188
NH_3 storage ($mtNH_3$)	0	10,600

also has significantly more PV capacity to provide more hours with enough energy to power both electrolysis and ammonia synthesis. This overbuilding of renewable capacity results in optimal curtailment of 18 % of generated wind and 24 % of generated PV, compared to only 1.5 % and 6.5 % in the flexible system.

Figure 6.3 shows a sample schedule for the flexible ammonia production facility for a week in July. Electrolysis operation generally follows wind energy availability because it is too energy intensive to be operated economically using a battery. Its annual average capacity (load) factor is 50 %, slightly higher than the capacity factor of 48 %

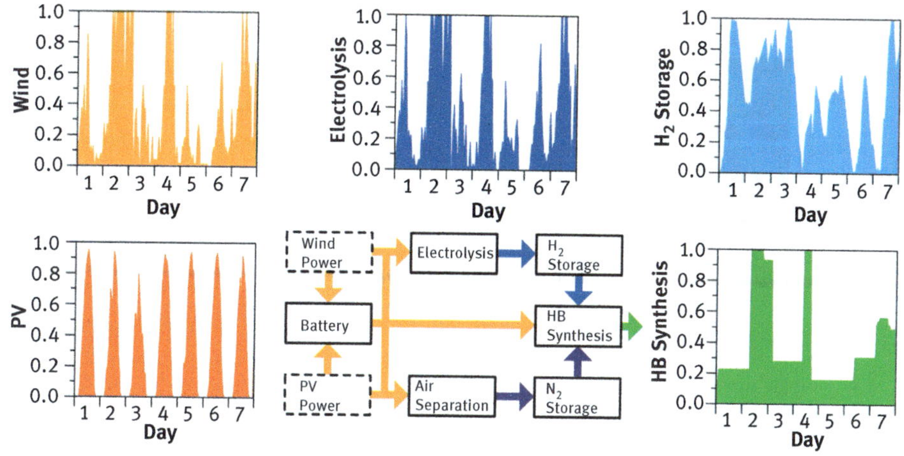

Figure 6.3: Optimal operational schedule for flexible renewable ammonia production facility for sample July week. The optimized hourly hydrogen and ammonia production rates as well as hydrogen storage inventory are shown in response to variable wind and solar generation profiles.

in this location. This demonstrates the benefit, although limited in this location, of combining wind and PV generation to improve the utilization of chemical production assets. Flexible ammonia synthesis changes its production rate on a multihour time scale, with a small hydrogen buffer storage used to decouple hydrogen and ammonia production rates over this interday timescale. Ammonia synthesis also has a minimum production rate of 10 % of installed capacity. The annual average capacity factor for ammonia synthesis is 54 %. Hydrogen storage is sized so that it could provide the minimum hydrogen demand of the synthesis process for approximately 30 hours. The air separation schedule is not included in Figure 6.3. It is slightly oversized, with an annual capacity factor of 85 %. Cryogenic nitrogen production uses a relatively low amount of energy, so the efficacy of oversizing to the same degree as that of electrolysis and ammonia synthesis is not worth the additional investment. Approximately 145 hours of nitrogen storage is included in the flexible optimized facility at minimum synthesis demand.

Figure 6.4 shows the effect of the production scale on LCOA, over a scale range of 10,000 mt/y to 500,000 mt/y. There are economies of scale; renewable ammonia is 20 % less expensive to produce on the scale of 500,000 mt/y than on the scale of 10,000 mt/y. This difference arises because of the lower capital cost for ammonia and nitrogen production on a larger scale. However, the investment costs of wind generation and electrolysis scale close to linearly. The scaling exponent for total capital investment is therefore close to 0.95 over the scale range considered. This is much closer to the linear capital cost scaling the 0.6–0.7 range for this exponent typically used in the chemical industry. However, noneconomic considerations are also important in the scaling of renewable ammonia facilities. The 500,000 mt/y facility would require 1,200 MW of wind generation and 1,150 MW of electrolysis. This wind capacity is close to six times that of the largest wind

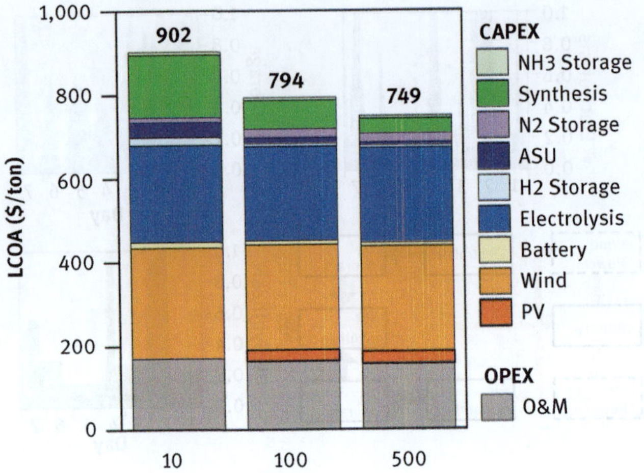

Figure 6.4: Levelized cost of ammonia for production scales of 10,000 mt/y, 100,000 mt/y, and 500,000 mt/y in Southwest Minnesota. All technology cost and performance data used to generate these results are listed in Chapter 5. O&M costs comprise annual maintenance costs as well as component replacement where required such as electrolysis stacks or ammonia synthesis catalyst.

farm currently operating in Minnesota. Meanwhile, at the end of 2022, only 366 MW of PEM electrolysis were installed globally [156].

Figure 6.5 shows the cost of ammonia production in five different locations in Minnesota that have factors of wind capacity ranging from 39 % to 49 %. Locations with higher wind capacity factors have lower LCOA. This occurs because less required over-

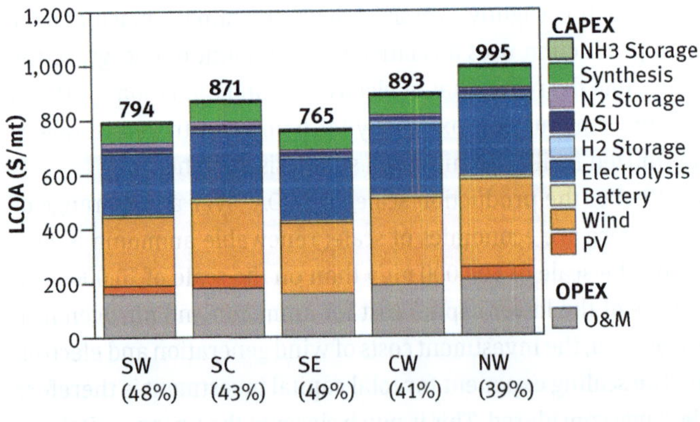

Figure 6.5: Levelized cost of renewable ammonia for 100,000 mt/y production scale in selected Minnesota locations with wind capacity factors ranging from 39 % to 49 %. The capacity factors for each location are listed in parenthesis below the location. All technology cost and performance data used to generate these results are listed in Chapter 5. O&M costs comprise annual maintenance costs as well as component replacement where required such as electrolysis stacks or ammonia synthesis catalyst.

sizing of most or all constituent technologies leads to lower capital investment. Smaller wind turbines can be installed to generate the same amount of energy in a year. More hours of available wind energy also enable smaller chemical production and storage operating at higher capacity factors. The electrolysis and synthesis capacity factors are each approximately 10 % higher for the lowest cost Southeast Minnesota location compared to the highest cost Northwest Minnesota location.

Figure 6.6 compares the cost of 100,000 mt/y ammonia production at the Southwest Minnesota location using present-day and 2030 technology costs. Considerable cost reductions are observed for wind generation and electrolysis. The required $/kW capital investment for both technologies is expected to decrease to 2030 and beyond, as described in Chapter 5. Furthermore, PEM electrolysis should become more efficient, which means that less renewable generation will be required on the same ammonia production scale. In this case study, the overall energy intensity for making renewable ammonia increases from 10.6 MWh/mt today to 9.4 MWh/mt in 2030. This enabled a 28 MW (12 %) decrease in the required wind capacity. Annual O&M costs, which are largely a function of total capital for each technology, also decrease due to both factors discussed above. In general, a reduction in LCOA of about 30 % is observed for this location.

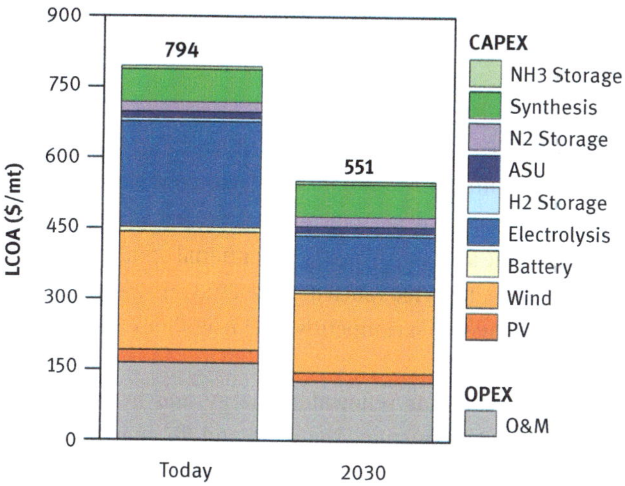

Figure 6.6: Levelized cost of renewable ammonia for present-day and 2030 at 100,000 mt/y production scale in Southwest Minnesota. All technology cost and performance data used to generate these results are listed in Chapter 5. O&M costs comprise annual maintenance costs as well as component replacement where required such as electrolysis stacks or ammonia synthesis catalyst.

Figure 6.7 contextualizes the results of this case study with respect to the current market for ammonia fertilizers. It specifically shows the average price of natural gas-based ammonia at the farmgate in Minnesota for 2011 through 2021, along with the present cost for renewable ammonia from the plant considered in southwest Minnesota.

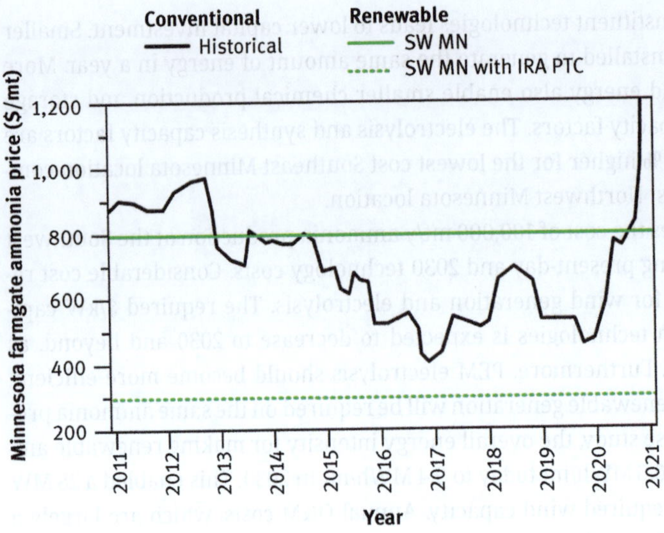

Figure 6.7: Comparison of modeled renewable ammonia production costs in Southwest Minnesota with and without production tax credits vs. the Minnesota farmgate market price of ammonia made with natural gas from 2011 to 2021. Production tax credits are legislated by the US Inflation Reduction Act of 2022 and apply to wind generation, PV generation, and hydrogen production. These credits are only received for 10 years of production but are levelized over a 25-year project lifetime.

Results are included for both nonsubsidized ammonia production considered in the previous figures of this analysis and including the economic benefit of federal production tax credits for wind energy, PV energy, and hydrogen [6]. Nonsubsidized ammonia production today is more expensive than the average price of conventional ammonia on the historical market. There have been certain periods such as 2011–2013 and 2020 and 2021 in which renewable ammonia would have been competitive today without any production tax credits.

The US federal production tax credits for renewable energy and hydrogen are transformative for renewable ammonia production. The wind and PV energy generation credits provide \$26.5/MWh [6]. The hydrogen production credits provide \$3/kg if the carbon intensity of the hydrogen is below 0.45kgCO_2/kgH_2 [6], as it certainly would be for 100 % wind-PV power sourcing. Combining these credits and converting to an ammonia basis, the present day system would be eligible for \$810/mt for the first 10 years of its production. Scaling these credits over a 25-year project lifetime using the 7.5 % WACC considered in this analysis gives a project-normalized credit value of \$498/mt. This scaled value is applied in the dashed line in Figure 6.7. There is a clear purely economic case for renewable ammonia due to the production tax credits. The net production cost is lower than the conventional market price for 2011 to 2021 and the production costs are also stable between each year.

6.2 Ammonia production from hydropower

Hydroelectric power generation varies on a much slower time scale than wind or solar-derived electricity. Its availability may exhibit seasonal variation, but stays close-to-constant in some locations. This means that the operational flexibility of an ammonia production facility is less crucial than for more intermittent resources. However, hydroelectric power generation also has a very high initial capital investment and thus higher energy production costs than wind and PV in regions with good potential. This consideration places a higher importance on the efficiency of power-to-ammonia. One approach might be hydrogen production via solid oxide electrolysis. Generally, the main drawback of SOEC is its limited operational flexibility relative to alkaline or PEM systems, but this is less of an issue with the minimally variable hydropower.

There have been a limited number of technoeconomic studies on using hydroelectric generation for ammonia production. Smith and Torrente-Murciano [349] analyzed hydropower-based ammonia production at various sites in Sierra Leone to provide fertilizer and to increase yields for domestic rice farming. They specifically surveyed 26 sites that had generation capacity ranging from 6 MW to 92 MW. This corresponds to ammonia production capacities of 4,000 to 70,000 mt/y. They found that domestic ammonia projects had a payback period of less than 5 years at 50 % of sites with the lowest specific hydroelectric $/MW cost relative to the business-as-usual practice of importing fertilizer. They found that the construction of hydroelectric dams contributed the highest proportion of the required investment cost.

Sousa et al. [352] found that a 25,000 mt/y hydropower-to-ammonia project in Norway had a present-day LCOA of $1,300/mt at an electricity price of $0.045/kWh. They also found an approximate LCOA dependence of $100/mt per $0.01/kWh change in electricity price. Their studied facility used a 30 MW PEM electrolysis system for hydrogen production with a capital cost of $1,100/kW. They identified that electricity cost is the economic driver of LCOA. They also outlined the potential to reduce the production cost to below $1,000/mt by 2030 as PEM electrolysis costs decrease and efficiencies improve.

Devkota et al. [79] performed a similar analysis for 410,000 mt/y hydropower-to-ammonia in Nepal. Their studied system used alkaline electrolysis with a capital cost of $425/kW. They found LCOA of $825/mt for an electricity cost of $50/MWh. They also performed a sensitivity analysis and found LCOA below $650/mt for energy costs below $35/MWh. According to the authors, such a decrease could reasonably be expected in the coming years as additional hydroelectric projects become operational in South Asia in the coming years.

In all three studies referenced above, energy was the largest contributor to the cost of making ammonia. As such, future technoeconomic studies should consider solid oxide electrolysis as a potential approach to improve facility efficiency and thus reduce energy consumption. The future cost of solid oxide electrolysis may be pivotal in determining the viability of the hydropower-to-ammonia concept. Of course, even if hydroelectric ammonia production is economically competitive, it will always be geographically constrained to areas with existing dams that can be powered or those where construction of the required generation infrastructure is appropriate.

6.3 Ammonia production from biomass

Ammonia production from biomass avoids electrolysis and the high associated energy intensity. Biomass is instead an alternative source of methane from which hydrogen can be produced following the conventional route of reforming followed by water gas shift and CO_2 removal. The use of biogenic carbon-based feedstocks reduces emissions intensity relative to newly extracted fossil fuels. Biomass-based ammonia has the potential for circularity in agriculture, whereby crop waste from one season is used to produce fertilizer for the next growing season of those same crops. Localized fertilizer production is a natural approach to achieve such circularity. Such an approach would be well suited to biomass specifically, which has a low value density and cannot be economically transported over long distances to centralized plants.

Technoeconomic analysis in the early to mid-2010s found biomass-to-ammonia production costs of at least \$600/mt. Tuna et al. [368] studied the production of ammonia from the steam reforming of anaerobic digestion biogas, as well as the gasification of woody biomass in Sweden. The largest production scales considered for the two approaches were 7,500 mt/y for biogas and 30,000 mt/y for biomass. This scale difference was attributed to the need for gasification technology to achieve economies of scale. They found that the resulting LCOA is approximately \$1,595/mt using biogas reforming and \$970/mt using biomass gasification. Biogas-based ammonia production had 25 % lower capital intensity despite its 4 times smaller scale. Its higher associated LCOA was due to the considerably higher cost of biogas feedstock, which was determined to be approximately 30 % higher than that of woody biomass. From an energy use point of view, the biogas reforming approach consumed 1.5 times that of natural gas-based ammonia production, while biomass gasification used twice as much as the conventional production route.

Andersson et al. [19] performed a similar analysis on ammonia production through gasification of wood biomass. They considered a production scale of 225,000 mt/y. They noted that a larger scale would likely have even better economics as a result of economies of scale on both biomass gasification and ammonia synthesis. They found an ammonia production cost of \$665/mt. Gasification and ammonia synthesis contributed 18 % and 45 %, respectively, to the total capital investment of approximately 560 MM\$. The main annual operating costs were the biomass feedstock and electrical power. The authors also considered heat and mass integration with a pulp and paper mill to improve economics. This integration reduced feedstock costs by 20 % by using waste bark from the pulp and paper mill to supplement wood, and improved energy efficiency by 10 %. In general, these improvements led to an ammonia production cost of \$585/mt.

Arora et al. [23] optimized the economics of the production of ammonia from biomass in three different countries. Specifically, they considered a 15,000 mt/y production scale using eucalyptus in Australia, straw in India, and bagasse in Brazil. They also compared autothermal reforming, steam reforming, and CO_2 reforming as alternatives to produce hydrogen from gasified biomass. They found that CO_2 reforming gave the lowest production costs in all considered locations. This reforming approach uses hot flue gas from biomass gasification to provide heat for the reforming reactions. This flue gas also acts as a nitrogen source, preventing the installation of a dedicated air separation unit for nitrogen production. Ultimately, they found that straw use in India gave the lowest ammonia production cost, which was approximately $680/mt. This resulted from India's low associated labor and energy costs. Bagasse-based ammonia production in Brazil had an associated LCOA of $950/mt. Brazil and India had similar labor and energy costs in this analysis, but bagasse gasification produces less syngas than straw, so the associated capital intensity is higher. Eucalyptus-based ammonia production had an LCOA of $1,250, mainly due to Australia's higher labor and energy costs. The cost of eucalyptus is also three to four times higher than that of straw or bagasse. The authors also performed a multiobjective optimization for each location to determine trade-offs between production cost and carbon intensity. They found that straw-based production resulted in the highest carbon intensity of approximately 1.5 $mtCO_2$/mt due to India's coal-dominate power generation mix. Using eucalyptus in Australia resulted in a carbon intensity around 1 $mtCO_2$/mt. Bagasse-based ammonia production had 0.5 $mtCO_2$/mt carbon intensity due to Brazil's high fraction of hydropower in their generation mix. For each location, the authors determined that using biomass for local onsite power generation could reduce carbon intensity, but would also increase LCOA due to the resulting lower ammonia production per unit feedstock input.

In the last 5 years, technoeconomic optimization efforts focusing on technology selection have resulted in LCOA below $450/mt in multiple scenarios. Zhang et al. [417] considered a 50,000 mt/y biomass-to-ammonia process, which used a high temperature entrained flow gasifier followed directly by high- and low-temperature water gas shift. The capital investment for the cost optimal facility was 171 MM$, with the highest cost contributions from the gasifier and heat exchanger network. The energy efficiency based on LHV was 45 % and the minimized production cost was $450/mt. The authors also optimized methane-based ammonia production on the same scale to provide a baseline for comparison, which had an efficiency of 61 % and a minimum production cost below $390/mt.

Sanchez et al. [326] compared four different technologies of syngas production, together with autothermal reforming vs. steam reforming technologies, and ammonia synthesis configurations of direct vs. indirect cooling for switchgrass ammonia production. The specific syngas generation processes were indirect gasification, direct gasification using pure oxygen and steam, direct gasification using air and steam, and anerobic digestion. A fixed switchgrass input of 600,000 mt/y was considered for all technology alternatives. This corresponds to ammonia production scales of 200,000 to

300,000 mt/y for gasification options and 50,000 mt/y from digestion. The resulting economics of digestion-based ammonia production was poor due to a lack of economies of scale. The best combination of technologies was determined to be indirect gasification, steam reforming, and direct reactor cooling. Each of these technologies had the lowest associated capital intensity among the alternatives for each function. In general, this configuration enabled an LCOA of \$425/mt, as well as the lowest carbon intensity of 0.75 mtCO_2/mt.

Demirhan et al. [75] optimized biomass-based ammonia production considering hardwood, corn stover, and municipal solid waste (MSW) as feedstocks. They determined production economics for three different US states with relevant feedstocks in each: Texas (hardwood and MSW), Iowa (corn stover and MSW), and California (hardwood and MSW). The largest and most economical production scale they analyzed was 330,000 mt/y and they optimized production facility designs to achieve 75 % reduction in GHG compared to current methane-based ammonia production. Overall, they found minimum LCOA using hardwood, corn stover, and MSW to be \$375/mt, \$492/mt, and \$457/mt, respectively. The production costs for hardwood and MSW-based ammonia were slightly lower in Texas due to its lowest industrial electricity costs. In California, optimally designed facilities burn biomass instead of purchasing more expensive grid electricity. This leads to 1 % and 5 % higher production costs than Texas for the use of hardwood and MSW, respectively. Iowa's MSW-based ammonia cost was 2 % higher than in Texas due to higher industrial electricity rates. Ammonia production costs from corn stover was they highest among the three feedstocks due to the higher unit price costs for this feedstock. The authors also determined the cost of ammonia production from natural gas using CO_2 capture and sequestration to achieve the same 75 % GHG emissions reductions. In general, these natural gas plus CCS production costs were higher than those of hardwood in Texas and California and MSW in Iowa and California.

In summary, ammonia produced from biomass can be economically competitive with conventional fossil-based ammonia using current-day technology. A key aspect to meaningful decarbonization using biomass-based ammonia will be careful tracking of associated emissions, as recent studies show a wide range of carbon intensities depending on the type of biomass used and the location in which this ammonia is produced. Furthermore, the question of how best to use a finite amount of available biomass remains unanswered. There is some argument to be made that these resources should be reserved for other "harder" to decarbonize applications such as aviation fuel that do not have a viable or well-developed electrified alternative.

7 Sustainable ammonia utilization for energy

This chapter discusses the economics of using ammonia in an energy context. The discussion specifically focuses on ammonia as (i) power and heat generation fuel, (ii) transportation fuel, and (iii) an energy carrier. It also includes a case study on using ammonia together with batteries and hydrogen in hybrid renewable energy storage systems.

7.1 Fuel for power and heat generation

Recent studies have assessed the cost of power generation using ammonia in a variety of technologies. Wang et al. [388] studied a 100 MW PtA system, which considered a reversible solid oxide electrolysis unit that could operate in both hydrogen production and hydrogen-to-power modes along with dedicated ammonia synthesis and decomposition reaction systems. Energy integration of the constituent processes in this system was pursued; exothermic heat from ammonia synthesis was used to heat steam for the SOFC in ammonia production or "charging" mode while the heat required for endothermic ammonia decomposition was obtained from the SOFC effluent in power generation or "discharging" mode. The optimal design of this system on the 100 MW scale allowed for overall efficiencies as high as 72 % and levelized energy cost as low as $240/MWh.

Sanchez et al. [323] determined optimal alternatives to the ammonia-to-power process. Potential constituent processes included a catalytic membrane reactor for ammonia decomposition, nitrogen/argon recovery to monetize these combustion inerts, gas turbines, selective catalytic reduction to remove NOx produced by ammonia combustion, and either secondary hydrogen combustion or recovery and recycle using a membrane. They determined ammonia-to-power efficiencies in the range of 30 to 45 % and LCOE as low as $230/MWh on the 100 MW scale. This minimum LCOE was achieved using membranes for both hydrogen recycle and nitrogen/argon recovery to achieve an additional revenue stream.

Cesaro et al. [54] also analyzed the economics of ammonia as fuel for combined cycle gas turbines on the 100 MW scale. They found that ammonia produced at $500/mt could give LCOE of $210/MWh for a power plant operating at a 25 % annual capacity factor and 70 %–30 % co-firing of hydrogen and ammonia. This LCOE could be reduced to $200/MWh if 100 % ammonia fueling is possible because this avoids additional investment in ammonia cracking. They found that by 2030, these costs would be comparable with natural gas-based power generation, which uses carbon capture and sequestration to achieve emissions intensity equivalent to that of ammonia-based power generation.

Blanco et al. [38] compared the generation of thermochemical and electrochemical power from ammonia from economic and efficiency perspectives. They found that a combined cycle gas turbine on the 100 MW scale could achieve an electricity generation cost of $290/MWh for an ammonia fuel price of $663/mt, with an ammonia-to-power efficiency of 34 %. Meanwhile, they found that a 1 MW anion exchange membrane (AEM)

https://doi.org/10.1515/9783110762525-007

fuel cell had an electricity generation cost of $1,200/MWh and an ammonia-to-power efficiency of 15 %. However, it is worth noting that the two-orders-of-magnitude lower generation scale and poor economics and efficiency of the electrochemical route are largely due to its relative lack of development at this time.

Richard et al. [310] also analyzed ammonia-fueled combined cycle gas turbines. They specifically compared two different ammonia decomposition technologies ("cracking") to facilitate a hydrogen-ammonia co-firing mixture containing 30 % hydrogen from a pure ammonia feedstock. The two ammonia decomposition technologies under consideration were fired tubular reactors and membrane reactors. They determined that membrane-based ammonia decomposition delivered the required amount of hydrogen with 30 % higher thermal efficiency and 10 % lower cost than a fired tubular reactor system. However, they also determined that the cost of ammonia was the main driver of the levelized cost of the energy obtained from the gas turbine. In general, they determined that both decomposition technologies led to a levelized energy cost in the range of $175/MWh to $225/MWh on the 400 MW scale using an ammonia fuel price of $475/mt.

Zhao et al. [423] analyzed the value of ammonia for seasonal energy storage specifically for use as a cofired fuel in coal-based power generation. They used an operational optimization model to compare a wind-PV-battery-coal energy system to one which could cofire up to 35 % renewable ammonia produced using excess renewable generation. They found that cofiring reduced annual operating costs by 10 %, annual emissions by 17 %, and renewable curtailment by 97 %.

Ammonia for long-duration energy storage in remote locations is a current use case of specific interest. These locations are often characterized by high conventional energy costs because fuels such as diesel need to be imported over long distances. Sagel et al. [317] determined the levelized cost of energy for a ammonia-based energy system for the Caribbean island of Curacao, which used 219 MW of wind generation. They specifically considered a small-scale, low-pressure ammonia synthesis process, which uses a ruthenium catalyst and absorbent to separate the ammonia product from the unreacted hydrogen and nitrogen. They determined that a 40,000 mt/y ammonia production system would be sufficient to balance monthly differences in wind generation and power demand, while Ni/Fe-redox batteries are used to balance short-term fluctuations. They consider SOFC technology for ammonia-to-power. Ultimately, the levelized cost of energy for this system was $130/MWh.

We previously analyzed renewable CHP systems in remote locations [298]. These locations are characterized by high conventional energy costs. We considered fully renewable energy systems that could produce hydrogen and ammonia and then use these chemicals for power generation with heat cogeneration or dedicated heat production in boilers. Optimally designed and operated systems for fully renewable CHP resulted in energy supply costs of $136.2/MWh and $90.6/MWh for Mahaka, Hawaii and the Northwest Arctic Borough, Alaska, respectively. These monthly costs are only 3 % more expensive than those that arise from historically low conventional power and fuel prices. Renewable ammonia was central to the economic viability of these systems. It was used as a power generation fuel with heat co-production in both locations. In the Northwest Arctic Borough, it was also used as a dedicated heating fuel. Its low storage cost allowed for seasonal heat storage, where ammonia was produced during the summer to be used in the winter.

Bounitis and Charitopoulos generalized the concept of ammonia-based power and heating decarbonization to the energy system of Great Britain [46]. Their economic modeling considered meeting power and heat demands in the 13 local distribution zones of Great Britain's current natural gas network over a full year of operation. For power generation, they considered the use of natural gas with carbon capture, biomass, nuclear, wind, and solar generation. Hydrogen and ammonia produced using electricity could also be used for power generation. For heating, they considered natural gas, hydrogen, and ammonia as potential fuels as well as the possibility of electrically driven heat pumps. Their modeling sought to determine the lowest-cost deployment of each of the above technologies while meeting a net zero emissions goal. In general, they found that the cost-optimal decarbonization pathway included between 2 and 6 TWh of ammonia storage depending on the specific scenario under investigation. This storage facilitated at minimum 80 % heating electrification and levelized cost of energy below \$85/MWh across all scenarios.

7.1.1 Levelized cost of energy from hydrogen, ammonia, and hybrid energy storage systems

Recent studies have compared hydrogen and ammonia as energy storage media and evaluated the benefits of using both together with batteries in hybrid energy storage systems (see Figure 7.1). In our previous work [295], we evaluated the economics of hydrogen- and ammonia-based energy storage on the 1 MW scale for 15 US cities, which comprehensively represent climate-demand regions throughout the continental US. We found that using hydrogen gave slightly lower LCOE in locations with high solar potential and/or less seasonal generation-demand mismatch, while ammonia provided up to 30 % reduction in LCOE in locations with high wind potential and larger seasonal storage requirements. We also found that combining the two storage media in hybrid energy storage systems gave the lowest LCOE in all locations. This LCOE was between \$0.17/kWh and \$0.28/kWh including investment in new renewable generation.

Sanchez et al. [325] performed a similar analysis for 14 different regions of Spain considering batteries, hydrogen, and ammonia available as energy storage media. They considered an annual average demand of 100 MW in each location. The lowest-cost fully renewable energy storage systems in each region used all three storage technologies in a hybrid configuration. This achieved the best overall trade-off between efficiency and storage cost in balancing intermittent wind and solar resources. The optimal levelized cost of energy was between \$82/MWh and \$105/MWh, depending on the region.

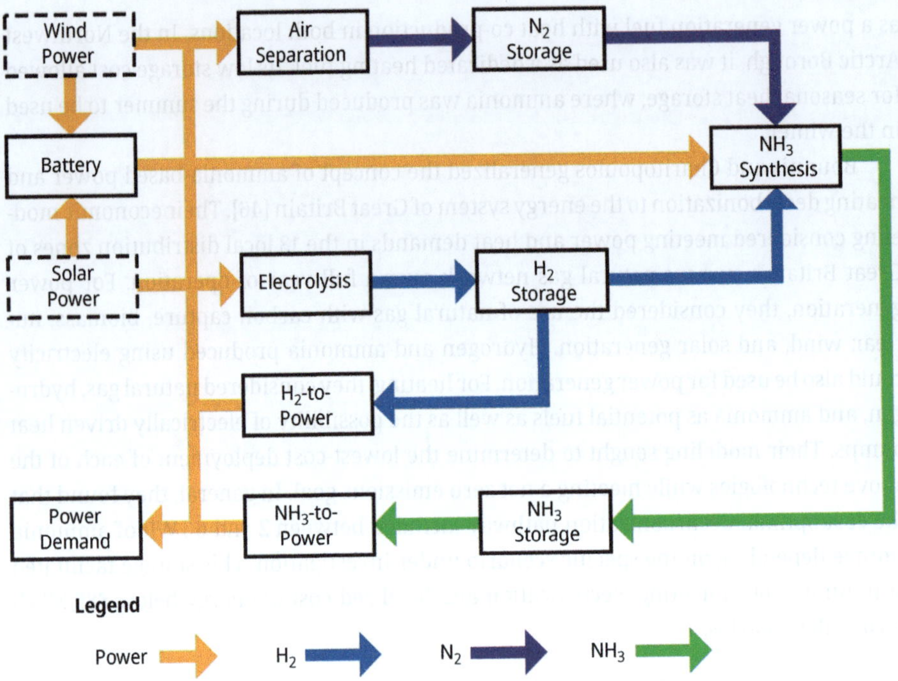

Figure 7.1: Conceptual flowsheet for hybrid renewable energy storage system including battery, hydrogen, and ammonia energy storage pathways. Electrical energy can be directly stored in batteries to balance generation and demand. This energy can also be stored by making hydrogen via electrolysis or further processing to ammonia by combining it with nitrogen. Both hydrogen and ammonia can be used to generate power through fuel cell and combustion technologies.

Mewafy et al. [253] performed optimal design modeling to analyze the economics and technology selection of microgrids that use wind and solar energy to supply power, heat, synthetic methane, and oxygen. These microgrids required electrolytic hydrogen production to facilitate synthetic methane production and byproduct oxygen supply. However, their optimization model also determined that the inclusion of renewable ammonia production allowed the lowest levelized cost of energy. This ammonia was used primarily to displace the demand for synthetic natural gas because ammonia is cheaper to store than the hydrogen required to make synthetic natural gas. Additionally, for this specific set of demands, the nitrogen production upstream of ammonia synthesis increased the overall oxygen production of the facility. This allowed the more energy-intensive electrolysis to operate more flexibly while still meeting overall oxygen demands.

Lan et al. [203] compared multiple energy conversion pathways involving hydrogen and ammonia to enable the supply of renewable energy. Specifically, they compared the production and use of hydrogen and/or ammonia in an area with a high renewable potential and transporting it via pipeline to an area of high energy demand versus high voltage power transmission to an area of high energy demand where hydrogen and/or ammonia production and utilization could facilitate stationary energy storage. For the power transmission-stationary storage approach, they also considered the possibility of hybrid hydrogen-ammonia energy storage. They found that this high-voltage transmission approach with hybrid storage results in the lowest levelized cost of energy of approximately \$80/MWh for a transmission distance of 100 km. High-voltage transmission was cheaper than pipeline transportation over this distance because of its higher overall efficiency. More specifically, pipeline transportation necessitates conversion of all generated power to hydrogen or ammonia, while direct electrical transmission only requires the conversion of excess power to "shift" it to later periods of renewable generation deficits. Meanwhile, ammonia's lower storage cost made its use optimal for longer-duration energy storage.

7.1.2 Case study: Ammonia for energy storage

This subsection contains a case study on the economics of using ammonia for renewable energy storage in addition to using hydrogen and batteries. This case study is adapted from our previous work [295], but considers a 100 MW annual average demand scale rather than the 1 MW systems analyzed previously. Specifically, we consider energy system configurations that use (i) batteries, (ii) batteries and hydrogen, and (iii) batteries, hydrogen, and ammonia. We also consider here the economic benefits of US federal production tax credits for wind energy, PV energy, and hydrogen [6]. The LCOE is minimized for systems in the same 15 cities considered in our previous work by simultaneously optimizing technology selection, sizing, and hourly operation. Publicly available hourly time-series data for renewable capacity factors [271] and power demand [268] in each location are used as input data. The optimization model constraints ensure that power demands are met in each hour, and all technologies operate within their specifications (e. g., turndown, ramp rates). For a more detailed description of this optimization model, we refer the reader to our previous work [295]. The technology cost projections for 2030 from Chapter 10 are used for renewable generation, batteries, electrolysis, PEM fuel cells for hydrogen-to-power and combined cycle gas turbines for ammonia-to-power. Cost data for 2030 is used because, at present, there is no need for long-duration storage as more renewable capacity must first be added to the energy infrastructure [332].

Figure 7.2 shows the optimal LCOE in each location for each configuration of the analyzed storage system. To contextualize these results, the estimated LCOE for new natural gas combined cycle gas turbine power plants is \$62/MWh or \$99/MWh if a \$30/mtCO_2 carbon sequestration cost is assumed [207]. Batteries are cost prohibitive as the sole means of energy storage in 100 % renewable systems. This is despite wind and PV production tax credits, the combined revenue of which ranges between \$15/MWh and \$16/MWh when normalized by power demand. Battery-alone systems only achieve LCOE below \$200/MWh in Miami, Phoenix, and Las Vegas, twice that of natural gas CCGT with CCS.

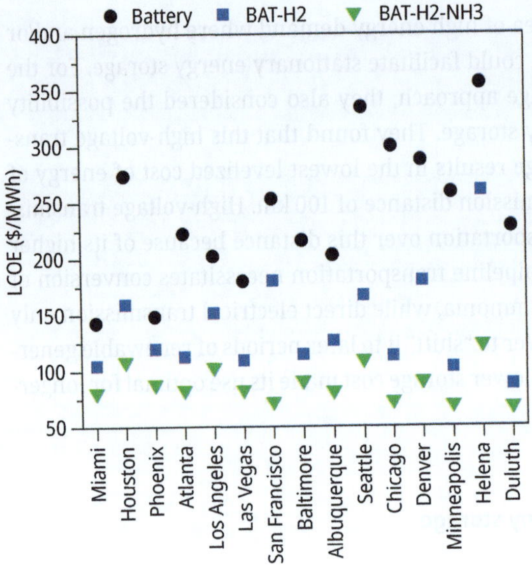

Figure 7.2: Optimal levelized cost of energy for battery, hybrid battery-hydrogen, and hybrid battery-hydrogen-ammonia energy storage systems, which use wind and solar generation to meet an annual average power demand of 100 MW. Results are shown for 15 cities, which represent all climate and moisture zones in the continental United States. All technology cost and performance data used to generate these results are listed in Chapter 5 and use 2030 projections where applicable.

These locations have relatively high PV potential and their periods of highest demand generally coincide with times of good PV generation. A major driver of power demand in these locations is air conditioning, which is most used during summer days.

Adding hydrogen as a "longer" duration energy storage medium improves economics at every location, with a reduction of up to 63 % LCOE in Chicago. Systems in 14 of the 15 locations have LCOE below $200/MWh with Helena being the only exception, while those in 8 of the 15 locations have LCOE below $120/MWh—within 20 % of NG CCGT with CCS. The Duluth battery-hydrogen system has an LCOE of $87/MWh. This is the only location where the economics is better than that of NG CCGT for the battery-hydrogen configuration. The improved economic performance of combined battery-hydrogen storage is largely facilitated by the lower $/MWh storage cost associated with hydrogen compared to a battery. These storage systems also enjoy the benefit of hydrogen production tax credits. Revenues from these credits range from $10/MWh to $30/MWh on demand-normalized basis; this brings combined credit revenues between $27/MWh and $52/MWh.

The addition of ammonia to create a hybrid system with all three storage media largely brings the economics of these renewable energy storage systems to parity, or better compared to NG CCGT with CSS. Specifically, these battery-hydrogen-ammonia systems have LCOE below $100/MWh in 11 of the 15 analyzed locations. The highest LCOE for this configuration is $122/MWh in Helena, which is a 53 % cost reduction compared to battery-hydrogen systems. Ammonia has much lower $/MWh storage costs than hydrogen, while its production still benefits from hydrogen production tax credits. For these hybrid systems, the total credit revenue is between $31/MWh and $51/MWh on a demand-normalized basis.

Table 7.1 shows the average optimal design of the battery, battery-hydrogen and battery-hydrogen-ammonia systems at the 15 locations. There exists a clear trend where the availability of lower-cost energy storage media (i. e., battery > hydrogen > ammonia) results in optimal systems having more total installed megawatt-hour of storage capacity and less renewable generation capacity. Specifically, a 3 times increase in storage capacity and 38 % reduction in renewable generation is observed when hydrogen can be used in combination with batteries. When ammonia can also be used, an additional 7 times increase in storage capacity and 30 % reduction in renewable generation is observed. This reduction in renewable generation capacity occurs despite the fact that batteries have the highest round-trip efficiency. In other words, when energy storage is expensive, it is more economical to overbuild renewable generation to directly meet a higher fraction of demand and then curtail excess generation. Thus, hybrid systems that use batteries, hydrogen, and ammonia can most effectively use generated renewable electricity in addition to providing the most economical competitor to conventional generation resources.

Table 7.1: Average optimal system design characteristics for renewable energy storage systems using batteries, batteries + H_2, and batteries + H_2 + NH_3.

Capacity	**Battery**	**BAT-H_2**	**BAT-H_2-NH_3**
Average LCOE ($/MWh)	241	139	88
Wind Turbines (MW)	395	385	285
PV Arrays (MW)	1,230	620	415
Battery (MWh)	3,685	1,120	810
PEM Electrolysis (MW)	0	190	160
H_2 Storage[1] (MWh)	0	8,065	970
PEM Fuel Cell (MW)	0	90	30
NH_3 Synthesis (mtNH_3/y)	0	0	85,000
NH_3 Storage[1] (MWh)	0	0	63,785
NH_3-fed CCGT (MW)	0	0	85

[1]Storage capacity in MWh based on conversion mass-to-power in H_2- or NH_3-to-power technologies.

Table 7.2 shows the specific optimal design for hybrid energy storage systems in Phoenix and Minneapolis. These cities were chosen for a more detailed examination for variety; Phoenix uses entirely solar generation, while Minneapolis uses the highest percentage of wind generation. The design motifs for the systems in these cities also help to provide a more global understanding of the trends in this study. An important aspect of

Table 7.2: Optimal LCOE and unit capacities for hybrid battery-H_2-NH_3 energy storage systems in Phoenix and Minneapolis.

Candidate Unit Capacity	Phoenix	Minneapolis
LCOE ($/MWh)	87.7	68.6
Wind Turbines (MW)	0	387
PV Arrays (MW)	626	141
Battery (MWh)	1,370	410
PEM Electrolysis (MW)	140 (*0.28*)	185 (*0.35*)
H_2 Storage (MWh)	448	1,730
PEM Fuel Cell (MW)	30 (*0.23*)	55 (*0.22*)
Air Separation (mtN_2/y)	28,000 (*0.72*)	39,500 (*0.74*)
N_2 Storage (mtN_2)	4,100	2,000
NH_3 Synthesis (mtNH_3/y)	58,000 (*0.43*)	90,000 (*0.40*)
NH_3 Storage (MWh)	45,325	47,780
NH_3-fed CCGT (MW)	74 (*0.09*)	82 (*0.12*)

[1] Capacity factors for chemical production units are italicized and in parenthesis after the installed capacity of chemical production units.

optimal energy storage system sizing is the annual capacity factors of chemical production and power generation technologies. These are the cumulative amounts of chemical production or power generated over a year divided by the maximum that could be produced or generated if operated at full capacity. These are shown in parentheses alongside the capacity for chemical production and power generation in Table 7.2.

In this study, the optimal electrolysis and ammonia synthesis capacity factors range between 29–42 % and 40–52 %, respectively. Electrolysis operates at lower capacity factors because it is more energy intensive; it is not economically feasible to operate it using a battery or another form of energy storage. Ammonia synthesis, on the other hand, requires less energy but must operate at least 10 % of its nameplate capacity. Optimal hydrogen-to-power and ammonia-to-power capacity factors are between 19–31 % and 6–20 %, respectively. In all systems, chemicals are made in more hours than they are used to generate power. This implies some value in the decoupling of "charge" and "discharge" infrastructure (e. g., electrolysis + ASU + synthesis vs. ammonia-fed CCGT) inherent to chemical energy storage. When hydrogen and ammonia forms of energy storage are compared, the higher capacity factor of hydrogen-based power generation coupled with the considerably higher amount of energy storage in the form of ammonia indicates that hydrogen-based power generation is actually used more frequently, while ammonia is used sparingly and stored over longer durations.

This observation is also supported by the optimal energy storage schedules for Minneapolis and Phoenix presented in Figure 7.3. These provide a helpful understanding of the interplay between hydrogen- and ammonia-based energy storage. In all investigated locations, ammonia is accumulated and used primarily on seasonal time scales.

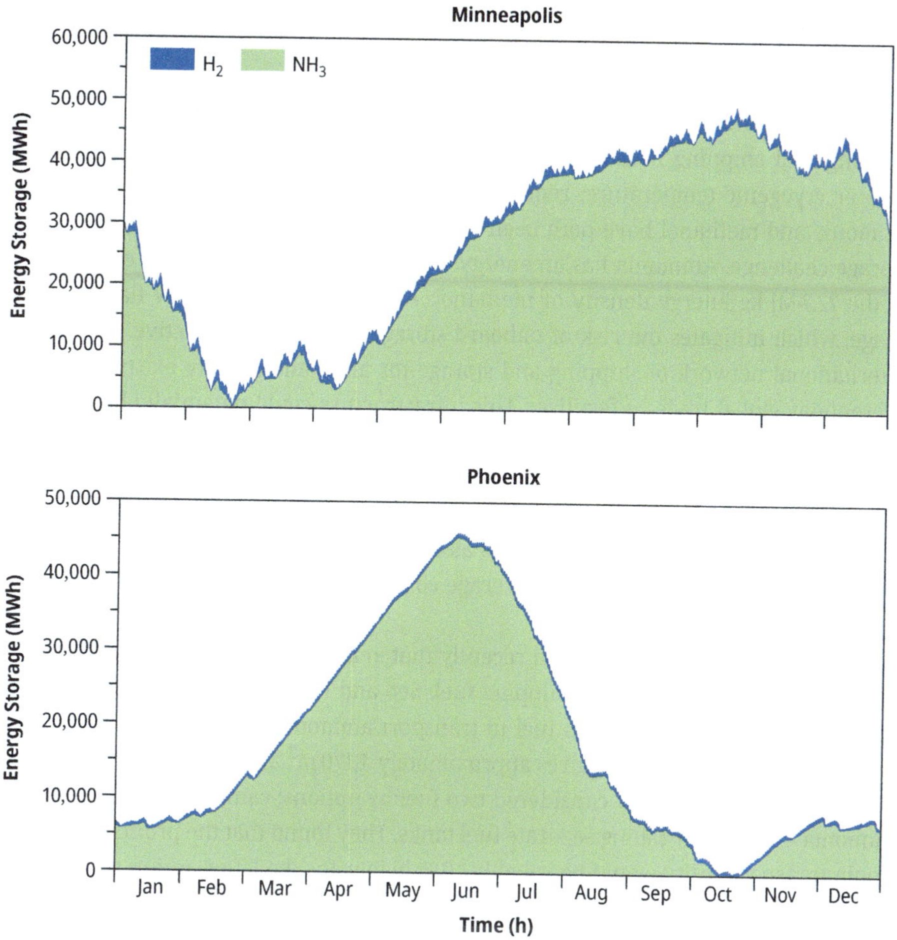

Figure 7.3: Optimal hydrogen and ammonia storage schedules for Minneapolis and Phoenix over a full year of system operation. The amount of energy stored in the form of hydrogen is stacked on top of the amount of energy stored in the form of ammonia. Storage inventories are listed in MWh units. Lower heating values (LHV) of hydrogen and ammonia are used to convert between mass and energy units.

In addition to that, hydrogen is used on shorter time scales, as evidenced by the considerable temporal fluctuations in the hydrogen storage schedules. The battery inventory is not shown in 7.3 because its relatively low capacity makes it almost imperceptible. Its storage fluctuates on an even faster timescale than that of hydrogen. For example, the battery for the Minneapolis system only has four hours of available capacity relative to average system demand. In summary, trade-offs between higher efficiency and lower storage cost can be exploited to reduce cost by using each energy storage medium for the appropriate purpose.

7.2 Ammonia as transportation fuel

Excitement around the potential of ammonia as a shipping fuel has been growing with interest in maritime decarbonization. Using hydrogen directly as fuel is not viable for long-haul shipping because of the cost and severe conditions, either high pressure or cryogenic temperature, required for its storage. The hydrogen-derived fuels ammonia and methanol have both been seriously considered because they avoid this storage challenge. Ammonia has an energy density of 22.5 MJ/kg, which is comparable to the 22.7 MJ/kg energy density of methanol. Ammonia also has a low flammability range, which mitigates the risk of onboard storage from that perspective. Finally, an international network of shipping and storage for ammonia already exists due to its present-day global trade as fertilizer. This infrastructure could potentially be used for fuel transportation, while related designs and experience could be easily transferred to new construction. For reference in the following discussion, two major conventional shipping fuels are heavy fuel oil (HFO) and marine gas oil (MGO). The global average production cost of HFO is \$7.9/GJ and its associated well-to-propeller emissions intensity is 87 $kgCO_2e$/GJ [17]. The global average cost of MGO is \$13.4/GJ and its emissions intensity is 87.8 $kgCO_2e$/GJ [17].

Some work has been published recently that analyzes the economics of the adoption of renewable ammonia as shipping fuel. Seo and Han [331] considered a specific case study of using ammonia as fuel to transport ammonia fertilizer from Kuwait to South Korea. This journey requires approximately 4,170 m^3 ammonia fuel to transport 84,000 m^3 of fertilizer. They considered two fueling options: cannibalizing some cargo ammonia as fuel or installing separate fuel tanks. They found that the profits of the ammonia transport entity would be roughly 10 % higher for the latter option wherein an additional fuel tank is installed to enable delivery of the full 84,000 m^3 of fertilizer.

Wu et al. [405] compared the feasibility of using ammonia-fed internal combustion engines and crackers with SOFC systems compared to conventional diesel and HFO combustion systems. They found that ammonia decomposition for use in hydrogen-fed SOFC would be more cost-competitive with conventional fuels than ammonia-fed ICE due to its higher fuel-to-power efficiency. However, considerable advances in volumetric power density and system lifespan must be achieved to maintain economic parity with HFO when green ammonia costs \$600/mt. However, the use of this cracker-SOFC system would also eliminate 87,000 $mtCO_2$/year for the 56 MW engine container ship that was considered in the case study in the analysis. The authors thus concluded that carbon tariffs could help accelerate the competitiveness of the ammonia cracking SOFC propulsion system.

Wang et al. [390] analyzed an ammonia-enabled sustainable future for global marine fuel using a modeling approach, which optimized a network of fuel production sites, transportation links, and fuel storage ("bunkering") ports. This analysis considered a 2050 scenario in which the production of 115,000,000 mt of green ammonia provides fuel for approximately 23 % of the global shipping trade. This considered fuel production quantity would require more than 1,000 GW of new solar generation and more than 100 GW of new wind generation. Ammonia production location decisions were heavily influenced by access to low-cost electricity. For example, no new production was located in Europe and shipping routes connected to Europe refueled at the Port of Aden in Yemen and Suez Port in Egypt. Ultimately, this analysis showed that ammonia fuel could optimally be produced and delivered to bunkering ports for $15.2/GJ ($341/mt).

Yang and Lam [409] studied the economic and operational aspects of a ship-to-ship ammonia bunker fuel supply chain. They compared pure ammonia fueling with the current use of MFO and also investigated dual MFO-ammonia fueling for a range of energy proportions. They found that pure ammonia bunkering could be up to 45 % less expensive than using MFO for an ammonia price of $300/mt. However, dual fueling strategies were determined to be at least as expensive as using MFO because of the longer fueling time that arose from sequential fueling. The main difference from MFO bunkering was an increased importance of fuel flow rate because of ammonia's lower energy content. The pure ammonia bunkering increased the fueling time by 60 %, while the dual fueling more than doubled it.

For ammonia to be successfully adopted as a low-carbon shipping fuel, a large increase in global renewable ammonia production will be required. For this reason, it is likely that multiple alternative fuels, including ammonia and methanol, will be used in the coming decades [127]. As many of these renewable fuels or "e-fuels" require hydrogen from electrolysis, the development of seawater electrolysis may be an appropriate priority in order not to stress freshwater reserves [338]. From a technology point of view, direct-fed ammonia fuel cells are likely to be more environmentally favorable than combustion in shipping applications. These fuel cells have the potential to be more efficient than combustion engines, resulting in less ammonia consumption for a given distance of travel. Furthermore, fuel cells could avoid combustion byproducts, such as nitrogen oxides. However, the development and cost of direct-fed ammonia fuel cells lag well behind those of combustion systems.

Ammonia has also been investigated as a rail freight fuel. As in shipping, its promise comes from low storage cost, relatively high energy density, and industrial experience in transporting ammonia via rail. Zhang et al. [421] compared three ammonia-based alternatives for rail engines: (i) decomposition ("cracker") with hydrogen-fed PEM fuel cell, (ii) cracker with hydrogen-fed solid oxide fuel cell (SOFC), and (iii) direct use in SOFC. Currently, the capital investment and replacement costs for the cracker-PEM system are considerably lower than those for any SOFC system. This is expected due to the relative stage of each of these systems' development. However, SOFC systems are more fuel efficient. In general, combining an ammonia cracker with a PEM fuel cell gave the lowest electricity cost among these ammonia-based alternatives for ammonia prices below $1,000/mt. To compete with diesel today without any carbon tax, renewable ammonia would need to cost below $400/mt for use in the cracker-PEM system. Meanwhile, SOFC systems would need a 70 % reduction in required capital investment to be competitive with diesel engines.

7.3 Ammonia as an energy carrier

Ammonia can be used to transport energy from areas of high renewable potential to areas of higher energy demand. Ammonia can be transported as a liquid via pipeline, rail, or ship; this is already done in the global trade of fertilizer. This energy can be "realized" in the form of electrical power or even hydrogen. Ammonia is sometimes referred to as a hydrogen carrier for this reason.

Tso et al. [366] compared ammonia with hydrogen as a means of storing and transporting energy from renewable generation sites to replace the demand for fossil energy in the five most populous cities in Texas. This analysis was based on supply chain optimization to determine the locations of renewable power generation, hydrogen production, and ammonia production to best meet energy demands. These chemicals were assumed to be transported via rail and truck. They found that for a storage duration of 1 month, using hydrogen gave an LCOE of $474/MWh. This was 40 % lower than the LCOE of ammonia as a result of the lower production cost of hydrogen and the higher roundtrip power-to-power efficiency. For a storage duration of three months, the use of ammonia and hydrogen resulted in LCOE of $758/MWh and $778/MWh respectively. The similar LCOEs for this storage duration demonstrate the trade-off between the above-described benefits of hydrogen and the lower storage cost of ammonia. Combining hydrogen and ammonia for a storage duration of 3 months reduced LCOE by 26 % to $556/MWh.

Demirhan et al. [74] proposed to connect energy facilities in regions with high renewable resources and high demand, considering liquefied hydrogen and ammonia to allow a low-cost interconnection. They conducted a case study analyzing the coupling of Texas and New York City (NYC), looking to replace 10 % of NYC's annual power demand. They compared the economics of local renewable generation in NYC and the import of energy carriers from Texas through the pipeline for both baseload and peak power consumption profiles. They found that using hydrogen and ammonia as an energy carrier reduced overall costs by up to 50 % compared to relying only on renewable generation in New York City. Specifically, hybrid wind PV systems in NYC had a LCOE of between $140/MWh and $150/MWh, while the transport of chemical energy reduced the LCOE to between $110/MWh and $116/MWh. For both baseload and peak demand profiles, the optimal ratio of imported hydrogen and ammonia was approximately 1:1 on an energy basis.

Sanchez et al. [324] investigated the merits of ammonia as an energy carrier to assist in the energy transition in Leon, Spain. They found that cost optimal systems with 75 % or more of total energy demand via renewable generation include ammonia production, transportation and utilization, and the power generation infrastructure. Ammonia production can be located in areas with the highest renewable potential, because it can be transported inexpensively and only requires air and water as feedstock.

Agyekum et al. [12] focused specifically on the cost transporting ammonia and other hydrogen carriers from a number of African countries to Port Blexan in Germany. They found that ammonia is the lowest cost hydrogen carrier among a number of alternatives, with transportation costs ranging from $0.19/kgH$_2$ for ammonia produced in Morocco to $0.55/kgH$_2$ for production in South Africa. The transportation cost of ammonia was at least 20 % lower on a per kgH$_2$ basis than methanol.

Sagel et al. [318] expanded on their previous analysis [317] to consider the possibility of importing ammonia to the small islands of Curacao and Fiji's largest island Viti Levu. They found that the benefit of producing ammonia on larger scales in facilities with high renewable potential could outweigh the transportation cost to these islands and resulted in lower levelized energy costs than local ammonia production. Specifically, renewable ammonia produced in Trinidad and Tobago or Brazil at a cost of $500/mt and imported to Curacao resulted in LCOE of $0.11/kWh, compared to their previously published $0.12/kWh associated with local ammonia production for energy storage. Meanwhile, ammonia produced at the same price in Australia or Chile and imported to Viti Levu resulted in LCOE of $0.37/kWh, while local ammonia production for energy storage resulted in LCOE above $1.10/kWh due to Viti Letu's comparably poor renewable potential.

Hank et al. [126] compared ammonia with other renewables-derived chemical energy carriers in terms of efficiency and cost. Specifically, they considered hydrogen-equivalent production of 42,500 mtH$_2$/y, which is 240,000 mtNH$_3$/y, in Morocco with annual average wind and PV capacity factors of 50 % and 27 %, respectively. They considered PEM electrolysis with capital cost of $660/kW and cavern hydrogen storage with capital cost of $12/kgH$_2$. They assumed a generic transportation distance of 4,000 km as well as the use of each energy carrier as fuel for its own transportation. For ammonia, approximately 1 % of the contained energy was determined to be consumed. In general, the combined cost of ammonia production and transport was lower than those of all other carriers studied. This was specifically due to its transportation cost being lower than that of liquified hydrogen or LOHC, and its more energy efficient production than that of methanol. The delivered cost of ammonia was approximately $135/MWh. From an efficiency at point of delivery standpoint, ammonia (47.7 %) was only less efficient than liquified hydrogen (52.4 %).

Verleysen et al. [380] compared two options for the production and utilization of renewable ammonia in the Belgian power sector: local production in Belgium versus Morocco production plus transport to Belgium. They performed this analysis through the lens of a design optimization model that included robustness against economic and performance uncertainty of the constituent technologies. Their analysis specifically compared a PV-based facility in Morocco with a 26 % annual average capacity factor to a facility in Belgium with a PV power in Belgium with a 13 % capacity factor and the ability to purchase grid power at $80/MWh. Both facilities used PEM electrolysis and hydrogen storage with nominal costs of \$950/kW and 890/kgH_2 respectively. They also considered the possibility of local backup power generation using a PEM fuel cell with a nominal cost of \$1,825/kW. For ammonia produced in Morocco, an ammonia-fueled ship was used for transportation to Belgium, consuming approximately 2 % of the transported cargo. They found that the production of ammonia in Morocco and its transport to Belgium resulted in ammonia supply costs below \$650/mt. On the other hand, ammonia produced in Belgium costs more than \$950/mt. They also found that uncertainties in the cost of renewable generation and electrolysis had the largest effect on ammonia production costs.

Egerer et al. [86] evaluated the economics of producing ammonia in Australia using wind and solar resources and subsequently transporting it to Germany for use in energy applications. The Australian location studied had 54.3 % and 28 % wind and PV capacity factors, each of which is at least 10 % higher than the German location of interest. In this study, 1,930,000 mt of ammonia was transported annually through the Suez Canal to Germany, over a distance of 17,600 km. They used 2030 cost projections to determine the cost of ammonia production, specifically \$440/kW for alkaline electrolysis, \$550/kW for PEM electrolysis, and \$500/kg for hydrogen storage. Their optimal ammonia production facility used both electrolysis technologies, taking advantage of the lower cost of alkaline and the higher efficiency and flexibility of PEM. They consider ammonia-fueled ships for the required international transportation; this shipping consumes roughly 3 % of the ammonia contained as fuel. Ultimately, the delivered cost of ammonia to Germany was determined to be approximately \$625/mt or \$120/MWh based on ammonia's lower heating value. They also found that direct use of this ammonia was preferential to cracking to hydrogen, as a delivered hydrogen cost of \$175/MWh was observed.

8 Ammonia for synergistic energy and agriculture use

The previous two chapters described the economics and other systems engineering aspects of sustainable ammonia production and utilization, respectively. The most straightforward application for sustainable ammonia production discussed in Chapter 6 is the use of this ammonia to reduce the carbon intensity of nitrogen fertilizer supply chains. In contrast, the potential applications for sustainable ammonia utilization discussed in Chapter 7 are as fuel for power and/or heat generation, as fuel for transportation, and as a carrier for hydrogen to link hydrogen production and utilization sites in different locations. There are also potential synergies in the use of sustainable ammonia to link agriculture and energy systems.

In addition to the long-standing need for nitrogen fertilizers, agriculture systems use fuels for a variety of purposes. Many modern farms use tractors that currently use diesel fuel. During the last decade, ammonia has been demonstrated as alternative tractor fuels at pilot site farms in Minnesota and Iowa, as discussed in Chapter 2. In addition, a variety of crop farming products, including wheat, corn, soybean, barley, rice, sorghum, sunflower seeds, rapeseed/canola, and oats, are dried after harvest to mitigate or prevent spoilage during storage caused by moisture content. Grain drying occurs over a relatively short period of time (e. g., days to a week) after harvest and requires a considerable amount of energy. Current grain drying burners typically use propane because it can be stored on site more easily than natural gas. Drying certain crops contributes meaningfully to the overall carbon intensity of their production. For corn farming, grain drying is responsible for more 40 % of total life cycle emissions [356]. Directly electrifying grain drying in an effort to reduce carbon intensity is difficult because of the relatively short time frame over which it occurs and the high energy demand required. Research at the University of Minnesota has explored the possibility of retrofitting grain drying burners to be fueled with ammonia. These experiments began in 2022 [32], and the operation of ammonia was publicly demonstrated at the Minnesota FarmFest in Morgan, Minnesota, in 2024. In general, ammonia has a potential future role in making agriculture more sustainable as a fertilizer and fuel.

Farms and their surrounding communities also have electricity demands. As discussed in Chapter 7, ammonia can be part of economical hybrid renewable energy storage systems by acting as long-term energy storage. Ammonia can also be used as a backup power generation fuel to provide reliability in the face of outages and other broader power system disruptions [312]. This is especially important for rural communities where many farms are located. These communities tend to have fewer redundancies in their electrical infrastructure and can sometimes even suffer from limited electricity access altogether. The potential use of ammonia as fertilizer, fuel and energy storage for combined agriculture-energy systems gives rise to the possibility of using the same chemical production, chemical storage, and energy storage technologies for

https://doi.org/10.1515/9783110762525-008

multiple end uses. For example, in hours with minimal renewable power generation, a battery can be used to meet electrical power demands on hourly timescales while also providing the minimum electrical power demands for nitrogen and ammonia production. As another example, ammonia will be removed from storage in the spring and/or fall for use as fertilizer. In contrast, seasonal energy generation demands in many locations will be highest in the summer due to increased air conditioning requirements and/or the winter due to increased heating requirements and decreased solar energy production. The potential benefit of using common system technologies is amplified by the different timescales on which electrical power, fertilizer, and fuel are needed. The responses of the system to renewable variability must be made instantaneously to meet the demands of power. Tractor fuel might be needed on a weekly or monthly basis, while fertilizer is needed once or twice a year, and grain drying fuel is only needed after harvest. This gives rise to the opportunity to flexibly produce ammonia as fertilizer and fuel during periods of surplus renewable generation. This can potentially lead to higher capacity factors for different production technologies, and thus better utilization of their required capital investment.

Some recent research has examined the above-described coupling of agriculture and energy systems enabled by sustainable ammonia. In our previous work, we conceptualized a farm-scale system that used only wind energy and biomass resources to meet power demands while also producing hydrogen and ammonia for use as fertilizer, tractor fuel, grain drying fuel, and energy storage for the West Central Research and Outreach Center (WCROC) of the University of Minnesota (UMM) and the adjacent University of Minnesota Morris campus (UMM) [297]. The potential energy storage technologies in this system included batteries, hydrogen fuel cells, and ammonia internal combustion (ICE) gensets (see Figure 8.1). These energy storage technologies also facilitate the export of excess generated power to the local utility to generate additional revenue for the farm energy system. We optimized the design of this system using a mixed-integer linear programming optimization formulation with hourly resolution optimal scheduling. A single representative week of synthetic data was used to represent each month of real historical data to reduce the computational time to solve this model. As in the results in Chapter 7, batteries, hydrogen, and ammonia were selected for energy storage to take advantage of the specific efficiency vs. storage cost characteristics of each. Furthermore, combining energy and agriculture demands led to an increase in hydrogen and ammonia production capacity factors, ultimately giving better utilization of capital investment and lower overall costs than considering energy and agriculture systems separately. A similar application of a farm-scale combined agriculture and energy system is discussed in more detail as a case study in the following Section 8.1.

Wang et al. [391] extended this system to include explicit management of nitrogen and water with the aim of decreasing nitrogen pollution in surrounding bodies of water. They also considered urea production from sustainable ammonia and captured CO_2 because urea is subject to less nitrogen volatilization than ammonia. They developed surrogate models to describe inherently nonlinear soil water and nitrogen dynamics,

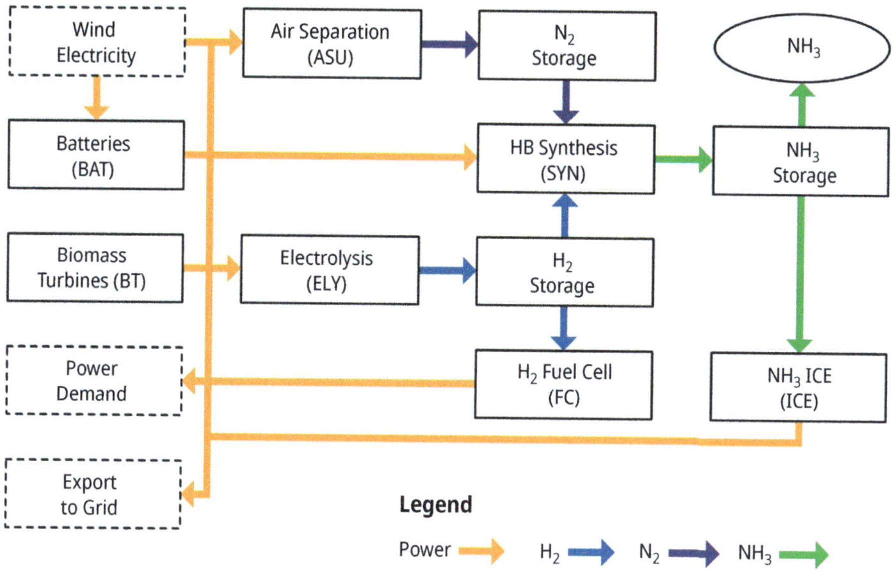

Figure 8.1: Conceptual flowsheet for combined agriculture and energy system enabled by sustainable ammonia. Power generated from wind resources is used to separate hydrogen from water via electrolysis, nitrogen from air via pressure swing adsorption, and to combine these feedstocks to make ammonia via Haber–Bosch synthesis. Ammonia must be made to meet fertilizer, tractor fuel, and grain drying fuel demands. Wind and solar energy can be stored in batteries to balance generation and demand while also enabling export of excess wind generation at favorable prices. This energy can also be stored in the form of hydrogen or ammonia, which can be used to generate power via fuel cells and internal combustion generator sets respectively. Biomass microturbines fed using crop waste can also be used for dispatchable power generation.

and embedded these in an optimal capacity planning and scheduling model in which input renewable generation, demand, and precipitation data was used to optimize the design of this system at the nexus of food-energy-water sustainability. They found that nitrogen pollution can be reduced by 57 % with only an increase in net present cost of 8 % by making urea from sustainable ammonia. The operation of fuel production and energy storage was virtually unchanged from the results in our previous work [297].

Edmonds et al. also considered sustainable ammonia production in an agricultural and energy context through the lens of microgrids in agricultural communities [85]. They considered a small-scale Haber–Bosch ammonia production system with a minimum production target of 5 tons per day and a direct ammonia-fed fuel cell embedded within the electricity distribution network of an agricultural community microgrid. Specifically, they analyzed the potential benefits of ammonia production located at one node of a 69 node electricity distribution system, which also includes distributed wind and solar generation, battery energy storage, dispatchable distributed generation fed with fossil fuels, and a grid connection. They developed an optimal scheduling model that coordinated the operation of these distributed energy resources and the ammonia plant as well as the three-phase power flows between each node in the distribution system. They conducted a number of different case studies to explore the benefits of this coordinated operation. First, they compared the implications of operating the ammonia plant as a constant load within the electricity system versus coordinating its operation as a demand response resource. They found that constant operation of the ammonia plant increased congestion, increased power loss and, therefore, higher location marginal prices (DLMP) within the distribution system. In contrast, coordinated operation of the ammonia plant reduces congestion and renewable curtailment by maximizing ammonia production during periods of high renewable generation. Consequently, the energy costs for ammonia production are approximately 33 % lower because DLMPs are reduced when renewable generation is more plentiful and congestion is alleviated. At the same time, economically optimal production results in 3.5 times more ammonia produced than during the constant operation case. The authors also considered an islanded mode, in which all demand must be met with distributed energy resources, and buying and selling power with the grid is not allowed. In this case, ammonia production can act as a controllable demand for excess renewable power, which would otherwise be curtailed because export is not allowed. This results in an additional 30 % more ammonia production. Finally, the authors considered an "emergency" islanded case wherein wind turbines are shut down and solar production is greatly reduced. In this case, the direct ammonia fuel cell provides essential power to the distribution system. In general, this work demonstrated that the coordinated operation of energy distribution and ammonia production has multiple potential benefits.

Ikaheimo et al. [147] considered a much larger spatial scale and examined the integration of the entire energy and agriculture sectors. They determined the optimal design of power, heat, and fertilizer supply networks in Germany, Finland, Sweden, Norway, Denmark, Estonia, Latvia, Lithuania, and Poland. Across these countries, the total power and heat demands considered in the model are 1,321 TWh and 474 TWh, respectively. For ammonia, 4.5 million mt/y are needed as feedstock for 3.7 million mt/y of nitrogen fertilizers consumed. The primary energy resources considered in the modeling included wind, solar, and hydropower, as well as biomass feedstocks. They used an optimal capacity planning model that selects and sizes the installation of new technologies in power, district heating, and ammonia production systems. They also included hourly resolution scheduling of each of these subsystems (and their interaction) over a year-long horizon. In their results, wind, solar and hydroelectric generation met most of the energy demands. However, the ability of ammonia to be stored and transported at low costs was critical in multiple ways. From an agricultural standpoint, Germany was the main fertilizer consumer of the countries considered in this analysis, but Norway, Sweden, and the Baltic countries (Estonia, Latvia, Lithuania) were the main producers. Additionally, over 15 million mt/y ammonia were used for heat and power generation, especially in Germany. Specifically, these results also showed shared use of ammonia pipelines for fertilizer and energy carrier purposes depending on the time of year, as fertilizer demands are seasonal. In this way, the synergies between agriculture and energy enabled by sustainable ammonia extend to these entire sectors rather than a specific facility.

8.1 Case study: Sustainable agriculture energy system in Morris, Minnesota

The case study described here uses a conceptual system design similar to that in Figure 8.1 as in our previous work [297] but considered more recent wind generation and power demand data, updated technology cost and performance projections, and a different pricing structure for the export of excess renewable energy. The combined system demands arise from the University of Minnesota's WCROC farm and adjacent UMM campus. The WCROC is home to two 1.65 MW wind turbines, which we assume can be used to (i) make fertilizer and fuel for the farm, (ii) power the UMM campus, and (iii) export excess power to the local utility at time-varying location marginal prices (LMPs). The farm has corn and soybean crops of 290 and 116 acres, respectively, which require a total of 40 mt/y of ammonia as fertilizer, tractor fuel, and grain drying fuel. Ammonia for fertilizer and tractor fuel is withdrawn from storage in April and October in 40 %/60 % and 35 %/65 % proportions, respectively. Ammonia for grain drying is only required in October. The UMM campus has an annual average power demand of 0.95 MW.

The cost optimized system has a net present cost of \$0.123MM/y when considering a 25-year project lifetime with a 10 % weighted average cost of capital. This arises from a capital cost of \$3.5MM and a net annual operating cost of -\$0.19MM/y. The operating cost is net negative due to the replacement of conventional fertilizer and fuel purchases, as well as a net export of an average of 0.65 MW of power (see Figure 8.2). The implementation of this system also avoids 4,325 mtCO_2/y by replacing the production of fertilizers and fuels that are currently imported, as well as the power generation replaced in the grid by the excess exported wind power. Ultimately, an emissions avoidance cost of approximately \$25/mtCO_2 is a promising result toward the efficacy of a combined agriculture-energy system enabled by sustainable ammonia.

The optimal size of each installed chemical production, storage and power generation technology is shown in Figure 8.2. Electrolysis, pressure swing adsorption (PSA), and ammonia synthesis all operate in a time-varying manner. Electrolysis has an optimal installed capacity of 275 kW with annual average capacity factor of 52 %. Both PSA and ammonia synthesis have annual average capacity factors of 85 % meaning they produce on average 0.27 mt/d and 0.3 mt/d based on their installed capacities. The time-varying operation of ammonia synthesis during certain periods of low wind power availability reduces the energy storage requirements of the system. Similar oversizing of PSA and a very small 0.1 mt nitrogen buffer storage offer allow for nitrogen to be produced at essentially the same stoichiometry-adjusted rate as ammonia synthesis. The low capital cost and energy intensity of PSA favor this design characteristic. The low electrolysis

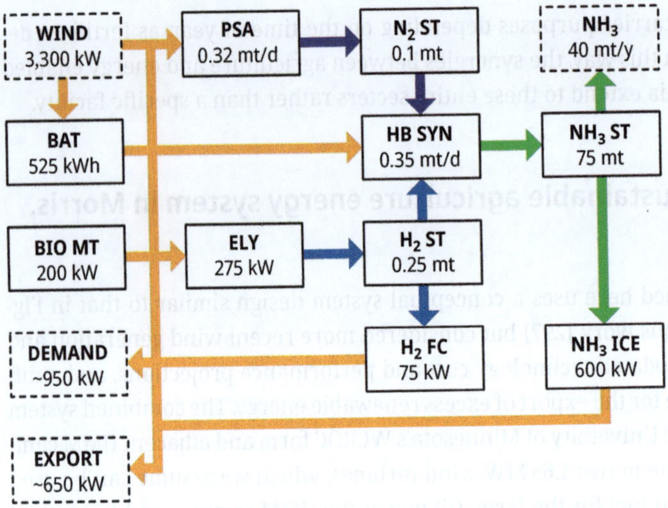

Figure 8.2: Optimal installed capacities for combined agriculture and energy system enabled by sustainable ammonia in Morris, Minnesota. Boxes with solid lines indicate installation of new technologies to facilitate the combined agriculture and energy system while boxes with dashed lines indicate energy or mass flows to or from existing infrastructure or demands. All technology cost and performance data used to generate these results are listed in Chapter 5.

capacity factor can be attributed to two factors. First, electrolysis is the most energy intensive of the three chemical production processes, so its unsteady operation offers the greatest benefit in terms of reducing energy consumption in periods of low wind power generation. Additionally, electrolysis has been oversized for the specific purpose of using hydrogen to store energy. The electrolysis capacity is approximately twice that dictated by the stoichiometry of the ammonia synthesis reaction if hydrogen were made only for the purpose of making ammonia.

Batteries, hydrogen fuel cells, ammonia-fed gensets, and microturbines fed with crop waste biomass are all included in the optimal system design as dispatchable sources of power generation. These generation pathways work in concert to provide the lowest-cost energy storage solutions. Batteries and hydrogen fuel cells are used for energy balancing over shorter timescales. They both have higher $/kWh storage costs than ammonia but higher roundtrip efficiency. In general, the power discharged from the 525 kWh battery represents 11 % of the total dispatchable power generation. The hydrogen fuel cell has an optimal capacity of 75 kW and an annual average capacity factor of 10 %. This represents 10 % of the total dispatchable generation. Ammonia is used for seasonal energy storage; it is only used to fuel power generation in June through September when wind generation is lower and UMM campus power demands are high. In general, the ammonia ICE gensets have an optimal capacity of 600 kW, a 7 % annual average capacity factor, and represent 35 % of the total dispatchable generation. Finally, biomass is used as the closest analogy to a conventional fossil fuel in the sense that it does not have

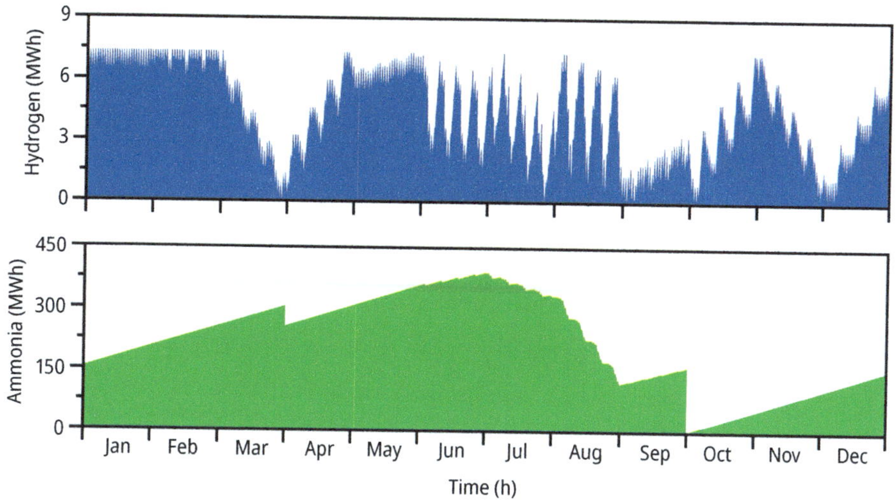

Figure 8.3: Optimal hydrogen and ammonia storage schedules for a full year of combined agriculture and energy system operation. Storage inventories are listed in MWh units. Lower heating values (LHV) of hydrogen and ammonia are used to convert between mass and energy units.

to be created by the agriculture energy system (as hydrogen and ammonia do) before it is used as fuel. It also has a minimal storage cost. Overall, 200 kW of biomass microturbines are used for 45 % of total dispatchable generation. Figure 8.3 shows the storage schedules for hydrogen and ammonia in kWh units. Two aspects of this figure make apparent the contrast between hydrogen and ammonia's role in the energy system. First, the 75 mt storage capacity of ammonia storage is approximately 50 times higher than that of hydrogen (0.25 mt) on an energy basis. The hydrogen storage inventory fluctuates much more than that of ammonia. This confirms that hydrogen is used for low-capacity, short-term energy storage, while ammonia is used for high-capacity, seasonal energy storage.

Bibliography

[1] Yara. Fertilizer Indsutry Handbook. Technical report, Yara, Oslo (Norway), 2018. yara.com/siteassets/investors/057-reports-and-presentations/other/2018/fertilizer-industry-handbook-2018-with-notes.pdf/.

[2] The Future of Hydrogen: Siezing today's opportunities. Technical report, International Energy Association, Paris (France), 2019. iea.org/reports/the-future-of-hydrogen.

[3] Ammonia: zero-carbon fertiliser, fuel and energy store - Policy briefing. Technical report, The Royal Society, London (United Kingdom), 2020. royalsociety.org/-/media/policy/projects/green-ammonia/green-ammonia-policy-briefing.pdf.

[4] World Energy Outlook 2020. Technical report, International Energy Association, Paris (France), 2020. iea.org/topics/world-energy-outlook.

[5] The ammonia market today and a bridge to the future. Technical report, Fertecon - S&P Global Commodity Insights, Washington DC (USA), 2024. ammoniaenergy.org/wp-content/uploads/2024/11/G-Eliseev-2024.pdf.

[6] 117th U.S. Congress. Inflation Reduction Act of 2022, H.R.5376. congress.gov/bill/117th-congress/house-bill/5376/text, 2022.

[7] AES Andes. AES Chile Submits Environmental Impact Assessment for Project Inna. aesandes.com/en/press-release/aes-andes-submits-environmental-impact-assessment-project-inna, 2024.

[8] Ahmed Afif, Nikdalila Radenahmad, Quentin Cheok, Shahriar Shams, Jung H Kim, and Abul K Azad. Ammonia-fed fuel cells: A comprehensive review. *Renewable and Sustainable Energy Reviews*, 60:822–835, 2016.

[9] Kodjo Agbossou, Mohanlal Kolhe, Jean Hamelin, and Tapan K Bose. Performance of a stand-alone renewable energy system based on energy storage as hydrogen. *IEEE Transactions on Energy Conversion*, 19(3):633–640, 2004.

[10] Babak Aghel, Sara Janati, Somchai Wongwises, and Mostafa Safdari Shadloo. Review on co2 capture by blended amine solutions. *International Journal of Greenhouse Gas Control*, 119:103715, 2022.

[11] Jennifer Aguinaldo. Neom hydrogen project reaches 60 % completion rate. meed.com/neom-hydrogen-project-reaches-60-completion-rate, 2024.

[12] Ephraim Bonah Agyekum, Jeffrey Dankwa Ampah, Solomon Eghosa Uhunamure, Karabo Shale, Ifeoma Prisca Onyenegecha, and Vladimir Ivanovich Velkin. Can africa serve europe with hydrogen energy from its renewables?—assessing the economics of shipping hydrogen and hydrogen carriers to europe from different parts of the continent. *Sustainability*, 15(8):6509, 2023.

[13] RK Ahluwalia, TQ Hua, JK Peng, and R Kumar. System level analysis of hydrogen storage options. *Proceedings of the 2008 US DOE Hydrogen Program Annual Merit Review*, 2010.

[14] Ken-ichi Aika. Role of alkali promoter in ammonia synthesis over ruthenium catalysts—effect on reaction mechanism. *Catalysis Today*, 286:14–20, 2017.

[15] Ken-Ichi Aika and Atsumu Ozaki. Kinetics and isotope effect of ammonia synthesis over ruthenium. *Journal of Catalysis*, 16(1):97–101, 1970.

[16] Aamer Akhand and Xiao-Yu Wu. Current research on gaseous ammonia detecting and capture technologies. *Current Opinion in Environmental Science & Health*, 36:100515, 2023.

[17] Fadhil Y Al-Aboosi, Mahmoud M El-Halwagi, Margaux Moore, and Rasmus B Nielsen. Renewable ammonia as an alternative fuel for the shipping industry. *Current Opinion in Chemical Engineering*, 31:100670, 2021.

[18] Andrew Allman, Prodromos Daoutidis, Douglas Tiffany, and Stephen Kelley. A framework for ammonia supply chain optimization incorporating conventional and renewable generation. *AIChE Journal*, 63(10):4390–4402, 2017.

[19] Jim Andersson and Joakim Lundgren. Techno-economic analysis of ammonia production via integrated biomass gasification. *Applied Energy*, 130:484–490, 2014.

https://doi.org/10.1515/9783110762525-009

[20] Oswaldo Andrés-Martínez, Mahdi Malmali, Qi Zhang, and Prodromos Daoutidis. Optimal design of an absorbent-enhanced ammonia synthesis process for solar thermochemical energy storage. *ACS Sustainable Chemistry & Engineering*, 12(25):9446–9460, 2024.

[21] Lori E Apodaca. Nitrogen (fixed) - Ammonia. U.S. Geological Survey, Mineral Commodity Summaries, January 2024. pubs.usgs.gov/periodicals/mcs2024/mcs2024-nitrogen.pdf, 2024.

[22] Julien Armijo and Cédric Philibert. Flexible production of green hydrogen and ammonia from variable solar and wind energy: Case study of chile and argentina. *International Journal of Hydrogen Energy*, 45(3):1541–1558, 2020.

[23] Pratham Arora, Andrew FA Hoadley, Sanjay M Mahajani, and Anuradda Ganesh. Multi-objective optimization of biomass based ammonia production-potential and perspective in different countries. *Journal of Cleaner Production*, 148:363–374, 2017.

[24] Kima Aswen. Project kima 2. kimaegypt.com/ProjectDetails.aspx?id=10, 2019.

[25] Australian Renewable Energy Hub. About the Australian Renewable Energy Hub. areh.com.au/about-AREH.html, 2024.

[26] Jonas Baltrusaitis. Editorial: Sustainable ammonia production. *ACS Sustainable Chemistry & Engineering*, 5(11):9257, 2017.

[27] R Bañares-Alcántara, G Dericks, M Fiaschetti, P Grünewald, J Masa Lopez, E Tsang, A Yang, L Ye, and S Zhao. Analysis of islanded ammonia-based energy storage systems. Technical report, University of Oxford, 2015.

[28] Jeffrey Bartels. A feasibility study of implementing an ammonia economy. Master's thesis, Iowa State University, 2008.

[29] Hervé Barthélémy, Mathilde Weber, and Françoise Barbier. Hydrogen storage: recent improvements and industrial perspectives. *International Journal of Hydrogen Energy*, 42(11):7254–7262, 2017.

[30] S Baruch-Mordo, JM Kiesecker, CM Kennedy, JR Oakleaf, and JJ Opperman. From paris to practice: sustainable implementation of renewable energy goals. *Environmental Research Letters*, 14(2):024013, 2019.

[31] Amanda Battersby. BP steps up to the plate on green hydrogen energy hub down under. upstreamonline.com/hydrogen/bp-steps-up-to-the-plate-on-green-hydrogen-energy-hub-down-under/2-1-1238457, 2020.

[32] Jeff Beach. Grain dryer experiment will put "green ammonia" to the test. agweek.com/news/minnesota-grain-dryer-experiment-will-put-green-ammonia-to-the-test, 2022.

[33] Joseph Beach, Jon Kinter, and Adam Welch. Rapid ramp reactor update. nh3fuelassociation.org/2018/12/08/rapid-ramp-nh3-prototype-reactor-update/, 2018.

[34] Luca Bertuccioli, Alvin Chan, David Hart, Franz Lehner, Ben Madden, and Eleanor Standen. Study on development of water electrolysis in the eu. https://www.fch.europa.eu/sites/default/files/study%20electrolyser_0.pdf, 2014.

[35] Ramesh Bhatia, RPS Malik, and Meera Bhatia. Direct and indirect economic impacts of the bhakra multipurpose dam, india. *Irrigation and Drainage: The Journal of the International Commission on Irrigation and Drainage*, 56(2-3):195–206, 2007.

[36] Abhoyjit Bhown and EL Cussler. Mechanism for selective ammonia transport through poly (vinylammonium thiocyanate) membranes. *Journal of the American Chemical Society*, 113(3):742–749, 1991.

[37] M Bielmann, UF Vogt, M Zimmermann, and A Züttel. Seasonal energy storage system based on hydrogen for self sufficient living. *Journal of Power Sources*, 196(8):4054–4060, 2011.

[38] Elena C Blanco, Antonio Sánchez, Mariano Martín, and Pastora Vega. Methanol and ammonia as emerging green fuels: Evaluation of a new power generation paradigm. *Renewable and Sustainable Energy Reviews*, 175:113195, 2023.

[39] Bloom Energy. Bloom Energy Celebrates Grand Opening of Fremont Multi-Gigawatt Factory, Adding Hundreds of New Clean Energy Jobs. bloomenergy.com/news/bloom-energy-celebrates-grand-opening-of-fremont-multi-gigawatt-factory-adding-hundreds-of-new-clean-energy-jobs/, 2022.

[40] Nick Bocker, Matthias Grahl, Akos Tota, Peter Häussinger, Paul Leitgeb, and Bernd Schmücker. *Nitrogen*, pages 1–27. John Wiley& Sons, Ltd, 2013. doi: https://doi.org/10.1002/14356007.a17_457.pub2.

[41] Alberto A Boretti. Novel heavy duty engine concept for operation dual fuel h2–nh3. *International Journal of Hydrogen Energy*, 37(9):7869–7876, 2012.

[42] Abhishek Bose, Nikifar Lazouski, Michal L Gala, Karthish Manthiram, and Dharik S Mallapragada. Spatial variation in cost of electricity-driven continuous ammonia production in the united states. *ACS Sustainable Chemistry & Engineering*, 10(24):7862–7872, 2022.

[43] Mirko R Bothien, Andrea Ciani, John P Wood, and Gerhard Fruechtel. Toward decarbonized power generation with gas turbines by using sequential combustion for burning hydrogen. *Journal of Engineering for Gas Turbines and Power*, 141(12):121013, 2019.

[44] Houssam Bouaboula, Mohammed Ouikhalfan, Ismael Saadoune, Jamal Chaouki, Abdelghafour Zaabout, and Youssef Belmabkhout. Addressing sustainable energy intermittence for green ammonia production. *Energy Reports*, 9:4507–4517, 2023.

[45] M Boudart. Ammonia synthesis: The bellwether reaction in heterogeneous catalysis. *Topics in Catalysis*, 1:405–414, 1994.

[46] Georgios L Bounitsis and Vassilis M Charitopoulos. The value of ammonia towards integrated power and heat system decarbonisation. *Sustainable Energy & Fuels*, 2024.

[47] John Brightling. Ammonia and the fertiliser industry: The development of ammonia at billingham. *Johnson Matthey Technology Review*, 62(1):32–47, 2018.

[48] David William Brubaker and Karl Kammermeyer. Separation of gases by plastic membranes-permeation rates and extent of separation. *Industrial and Engineering Chemistry*, 46(4):733–739, 1954.

[49] W Buechner, R Schliebs, G Winter, and KH Buechel. *Industrial inorganic chemistry. Industrielle anorganische Chemie*. Bayer A.G., 1986.

[50] Nicolas Campion, Hossein Nami, Philip R Swisher, Peter Vang Hendriksen, and Marie Münster. Techno-economic assessment of green ammonia production with different wind and solar potentials. *Renewable and Sustainable Energy Reviews*, 173:113057, 2023.

[51] Olivier Camus, Semali Perera, Barry Crittenden, Yvonne C van Delft, Dick F Meyer, Paul PAC Pex, Izumi Kumakiri, Sylvain Miachon, Jean-Alain Dalmon, Steve Tennison, et al. Ceramic membranes for ammonia recovery. *AIChE Journal*, 52(6):2055–2065, 2006.

[52] Alessandro Cappelletti and Francesco Martelli. Investigation of a pure hydrogen fueled gas turbine burner. *International Journal of Hydrogen Energy*, 42(15):10513–10523, 2017.

[53] Marcelo Carmo, David L Fritz, Jürgen Mergel, and Detlef Stolten. A comprehensive review on pem water electrolysis. *International Journal of Hydrogen Energy*, 38(12):4901–4934, 2013.

[54] Zac Cesaro, Matthew Ives, Richard Nayak-Luke, Mike Mason, and René Bañares-Alcántara. Ammonia to power: Forecasting the levelized cost of electricity from green ammonia in large-scale power plants. *Applied Energy*, 282:116009, 2021.

[55] Izzat Iqbal Cheema and Ulrike Krewer. Operating envelope of haber–bosch process design for power-to-ammonia. *RSC Advances*, 8(61):34926–34936, 2018.

[56] Ping Chen, Zhitao Xiong, Jizhong Luo, Jianyi Lin, and Kuang Lee Tan. Interaction between lithium amide and lithium hydride. *The Journal of Physical Chemistry B*, 107(39):10967–10970, 2003.

[57] Yifeng Chen, Bin Zhang, Yi Su, Chunjie Sui, and Junqing Zhang. Effect and mechanism of combustion enhancement and emission reduction for non-premixed pure ammonia combustion based on fuel preheating. *Fuel*, 308:122017, 2022.

[58] Ning-Ning Cheng, Zi-Liang Li, Hong-Chao Lan, Wen-Long Xu, Wen-Jing Jiang, Kuan Huang, and Hai-Long Peng. Deep eutectic solvents with multiple weak acid sites for highly efficient, reversible and selective absorption of ammonia. *Separation and Purification Technology*, 269:118791, 2021.

[59] ON Chhabra. Replacement of electrolysis plant with steam naphtha reformation plant, its integration with old heavy water plant (Paper No. 2.6). inis.iaea.org/collection/NCLCollectionStore/_Public/24/053/24053969.pdf, 1991.

[60] Kanghee Cho, Hyung Chul Yoon, Hee Tae Beum, Sun Hyung Kim, Chan Hyun Lee, and Jong-Nam Kim. Facile synthesis of carbon-based nanoporous adsorbent exhibiting high ammonia uptake under low pressure range. *Microporous and Mesoporous Materials*, 307:110460, 2020.

[61] Claus Hviid Christensen, Tue Johannessen, Rasmus Zink Sørensen, and Jens K Nørskov. Towards an ammonia-mediated hydrogen economy? *Catalysis Today*, 111(1-2):140–144, 2006.

[62] Core Writing Team, H. Lee and J. Romero (eds.). *Climate Change 2023: Synthesis Report. Contribution of Working Groups I, II and III to the Sixth Assessment Report of the Intergovernmental Panel on Climate Change*. IPCC (Geneva, Switzerland), 2023.

[63] Core Writing Team, Pachauri, R. K and Reisinger, A. (eds.). *Climate Change 2007: Synthesis Report. Contribution of Working Groups I, II and III to the Fourth Assessment Report of the Intergovernmental Panel on Climate Change*. IPCC (Geneva, Switzerland), 2007.

[64] Core Writing Team, R. K. Pachauri and L. A. Meyer (eds.). *Climate Change 2014: Synthesis Report. Contribution of Working Groups I, II and III to the Fifth Assessment Report of the Intergovernmental Panel on Climate Change*. IPCC (Geneva, Switzerland), 2014.

[65] JGC Holdings Corportation. News Releases 2023: Progress in Demonstrating Technology for Green Ammonia Production. jgc.com/en/news/2023/20230322-3.html, 2023.

[66] Edward Cussler, Alon McCormick, Michael Reese, and Mahdi Malmali. Ammonia synthesis at low pressure. *Journal of Visualized Experiments: JoVE*, (126):55691, 2017.

[67] Aldo Vieira Da Rosa and Juan Carlos Ordonez. *Fundamentals of renewable energy processes*. Academic Press, 2021.

[68] A Daisley and JSJ Hargreaves. Metal nitrides, the mars-van krevelen mechanism and heterogeneously catalysed ammonia synthesis. *Catalysis Today*, 423:113874, 2023.

[69] Aswath Damodaran. Cost of Equity and Capital (US). pages.stern.nyu.edu/adamodar/New_Home_Page/datafile/wacc.html, 2023.

[70] William IF David, Joshua W Makepeace, Samantha K Callear, Hazel MA Hunter, James D Taylor, Thomas J Wood, and Martin O Jones. Hydrogen production from ammonia using sodium amide. *Journal of the American Chemical Society*, 136(38):13082–13085, 2014.

[71] Arnout de Pee, Dickon Pinner, Occo Roelofsen, Ken Somers, Eveline Speelman, and Maaike Witteveen. Decarbonization of industrial sectors: the next frontier. Technical report, McKinsey & Company, New York, NY (United States), 2018. mckinsey.com/business-functions/sustainability/our-insights/how-industry-can-move-toward-a-low-carbon-future.

[72] Armando de Rezende, Mahdi Malmali, Pavlo O Dral, Hans Lischka, Daniel Tunega, and Adelia JA Aquino. Machine learning for designing mixed metal halides for efficient ammonia separation and storage. *The Journal of Physical Chemistry C*, 126(29):12184–12196, 2022.

[73] Jason Deign. Siemens Tests Ammonia as a Form of Energy Storage for Renewables. greentechmedia.com/articles/read/siemens-ammonia-hydrogen-energy-storage, 2018.

[74] C Doga Demirhan, William W Tso, Joseph B Powell, Clara F Heuberger, and Efstratios N Pistikopoulos. A multiscale energy systems engineering approach for renewable power generation and storage optimization. *Industrial & Engineering Chemistry Research*, 59(16):7706–7721, 2020.

[75] C Doga Demirhan, William W Tso, Joseph B Powell, and Efstratios N Pistikopoulos. Sustainable ammonia production through process synthesis and global optimization. *AIChE Journal*, 65(7), 2019.

[76] Dongshun Deng, Xiuzhi Duan, Bao Gao, Chao Zhang, Xiaoxia Deng, and Lei Gong. Efficient and reversible absorption of nh 3 by functional azole–glycerol deep eutectic solvents. *New Journal of Chemistry*, 43(29):11636–11642, 2019.

[77] Dongshun Deng, Bao Gao, Chao Zhang, Xiuzhi Duan, Yanhong Cui, and Jingheng Ning. Investigation of protic nh4scn-based deep eutectic solvents as highly efficient and reversible nh3 absorbents. *Chemical Engineering Journal*, 358:936–943, 2019.

[78] Xiaoxia Deng, Xiuzhi Duan, Lei Gong, and Dongshun Deng. Ammonia solubility, density, and viscosity of choline chloride–dihydric alcohol deep eutectic solvents. *Journal of Chemical and Engineering Data*, 65(10):4845–4854, 2020.

[79] Sijan Devkota, Sagar Ban, Rakesh Shrestha, and Bibek Uprety. Techno-economic analysis of hydropower based green ammonia plant for urea production in nepal. *International Journal of Hydrogen Energy*, 2023.

[80] Francisco Díaz-González, Andreas Sumper, Oriol Gomis-Bellmunt, and Roberto Villafáfila-Robles. A review of energy storage technologies for wind power applications. *Renewable and Sustainable Energy Reviews*, 16(4):2154–2171, 2012.

[81] Tharaka Rama Krishna C Doddapaneni, Ramasamy Praveenkumar, Henrik Tolvanen, Jukka Rintala, and Jukka Konttinen. Techno-economic evaluation of integrating torrefaction with anaerobic digestion. *Applied Energy*, 213:272–284, 2018.

[82] Xiuzhi Duan, Yanhong Cui, Chao Zhang, Bao Gao, and Dongshun Deng. Solubilities and thermodynamic properties of nh3 in glycerin and its derivatives. *Journal of Chemical and Engineering Data*, 64(3):1131–1139, 2019.

[83] Xiuzhi Duan, Bao Gao, Chao Zhang, and Dongshun Deng. Solubility and thermodynamic properties of nh3 in choline chloride-based deep eutectic solvents. *Journal of Chemical Thermodynamics*, 133:79–84, 2019.

[84] Xuekui Duan, Donghun Kim, Katabathini Narasimharao, Shaeel Al-Thabaiti, and Michael Tsapatsis. High-performance ammonia-selective mfi nanosheet membranes. *Chemical Communications*, 57(5):580–582, 2021.

[85] Lawryn Edmonds, Peter Pfromm, Vincent Amanor-Boadu, Mary Hill, and Hongyu Wu. Green ammonia production-enabled demand flexibility in agricultural community microgrids with distributed renewables. *Sustainable Energy, Grids and Networks*, 31:100736, 2022.

[86] Jonas Egerer, Veronika Grimm, Kiana Niazmand, and Philipp Runge. The economics of global green ammonia trade–"shipping australian wind and sunshine to germany". *Applied Energy*, 334:120662, 2023.

[87] Nico Eigen, Claude Keller, Martin Dornheim, Thomas Klassen, and Rüdiger Bormann. Industrial production of light metal hydrides for hydrogen storage. *Scripta Materialia*, 56(10):847–851, 2007.

[88] Carolyn C Elam, Catherine E Gregoire Padró, Gary Sandrock, Andreas Luzzi, Peter Lindblad, and Elisabet Fjermestad Hagen. Realizing the hydrogen future: The International Energy Agency's efforts to advance hydrogen energy technologies. *International Journal of Hydrogen Energy*, 28(6):601–607, 2003.

[89] Enaex. Enaex peru history. www.enaex.com/pe/us/enaex-peru-history/, 2020.

[90] Nicolaas Engelbrecht, Steven Chiuta, and Dmitri G Bessarabov. A highly efficient autothermal microchannel reactor for ammonia decomposition: Analysis of hydrogen production in transient and steady-state regimes. *Journal of Power Sources*, 386:47–55, 2018.

[91] Jan Willem Erisman, Mark A Sutton, James Galloway, Zbigniew Klimont, and Wilfried Winiwarter. How a century of ammonia synthesis changed the world. *Nature Geoscience*, 1(10):636–639, 2008.

[92] FA Ernst and MS Sherman. The world's inorganic nitrogen industry. *Industrial and Engineering Chemistry*, 19(2):196–204, 1927.

[93] Marco R Errera, TA da C Dias, Diego Mauricio Yepes Maya, and Electo Eduardo Silva Lora. Global bioenergy potentials projections for 2050. *Biomass & Bioenergy*, 170:106721, 2023.

[94] G Ertl, D Prigge, R Schloegl, and M Weiss. Surface characterization of ammonia synthesis catalysts. *Journal of Catalysis*, 79(2):359–377, 1983.

[95] Dan Esposito. Hydrogen Policy's Narrow Path: Delusions and Solutions. energyinnovation.org/wp-content/uploads/Hydrogen-Policys-Narrow-Path-Delusions-and-Solutions-2.pdf, 2024.

[96] Nand K Fageria and Virupax C Baligar. Enhancing nitrogen use efficiency in crop plants. *Advances in Agronomy*, 88:97–185, 2005.

[97] Mahdi Fasihi, Robert Weiss, Jouni Savolainen, and Christian Breyer. Global potential of green ammonia based on hybrid pv-wind power plants. *Applied Energy*, page 116170, 2021.

[98] Carlos A Fernandez, Oliver Chapman, Marilyn A Brown, Christian E Alvarez-Pugliese, and Marta C Hatzell. Achieving decentralized, electrified, and decarbonized ammonia production. *Environmental Science & Technology*, 58(16):6964–6977, 2024.

[99] Julian Florez, Mohammed AlAbbad, Holkan Vazquez-Sanchez, Miguel Gonzalez Morales, and S Mani Sarathy. Optimizing islanded green ammonia and hydrogen production and export from saudi arabia. *International Journal of Hydrogen Energy*, 56:959–972, 2024.

[100] Food and Agriculture Organization of the United Nations. Land, Inputs, and Sustainability: Fertilizers by Nutrient. fao.org/faostat/en/#data/RFN, 2024.

[101] Fortescue Energy. Our Green Projects: Gibson Island. energy.fortescue.com/en/our-green-projects/gibson-island, 2022.

[102] Fortescue Energy. CF Industries and POET to Demonstrate the Use of Low-Carbon Fertilizer in Corn Production to Reduce Carbon Intensity of Ethanol. cfindustries.com/newsroom/2024/cf-poet-low-carbon-ammonia, 2024.

[103] Stefano Frigo and Roberto Gentili. Analysis of the behaviour of a 4-stroke si engine fuelled with ammonia and hydrogen. *International Journal of Hydrogen Energy*, 38(3):1607–1615, 2013.

[104] Ulrik Frohlke. World's first of its kind green ammonia plant inaugurated by Skovgaard Energy, Vestas, and Topsoe. topsoe.com/press-releases/worlds-first-of-its-kind-green-ammonia-plant-inaugurated-by-skovgaard-energy-vestas-and-topsoe, 2024.

[105] Yasushi Fujimuri et al. Demonstration and Optimization of Green Ammonia Production Operation Responding to Fluctuating Hydrogen Production from Renewable Energy. ammoniaenergy.org/presentations/demonstration-and-optimization-of-green-ammonia-production-operation-responding-to-fluctuating-hydrogen-production-from-renewable-energy/, 2019.

[106] Gerda Gahleitner. Hydrogen from renewable electricity: An international review of power-to-gas pilot plants for stationary applications. *International Journal of Hydrogen Energy*, 38(5):2039–2061, 2013.

[107] Emre Gençer, Gary R Burniske, Otto C Doering III, Wallace E Tyner, Rakesh Agrawal, W Nicholas Delgass, Gebisa Ejeta, Maureen C McCann, and Nicholas C Carpita. Sustainable production of ammonia fertilizers from biomass. *Biofuels, Bioproducts and Biorefining*, 14(4):725–733, 2020.

[108] Giovanni Genova and Sergio Panza. Casale flexible green ammonia plant, the economically viable green production. ureaknowhow.com/2022-06-genova-casale-flexible-green-ammonia-plant-the-economically-viable-green-production/, 2022.

[109] Seyedehhoma Ghavam, Maria Vahdati, IA Wilson, and Peter Styring. Sustainable ammonia production processes. *Frontiers in Energy Research*, 9:34, 2021.

[110] Jeff Goldmeer (GE Power). Power to gas: Hydrogen for power generation. ge.com/content/dam/gepower/global/en_US/documents/fuel-flexibility/GEA33861%20Power%20to%20Gas%20-%20Hydrogen%20for%20Power%20Generation.pdf, 2019.

[111] Maraisa Gonçalves, Laura Sánchez-García, Erika de Oliveira Jardim, Joaquín Silvestre-Albero, and Francisco Rodriguez-Reinoso. Ammonia removal using activated carbons: effect of the surface chemistry in dry and moist conditions. *Environmental Science & Technology*, 45(24):10605–10610, 2011.

[112] Will Gorman, Julie Mulvaney Kemp, Joseph Rand, Joachim Seel, Ryan Wiser, Nick Manderlink, Fredrich Kahrl, Kevin Porter, and Will Cotton. Grid connection barriers to renewable energy deployment in the united states. *Joule*, 9(2), 2025.

[113] Matthew J. Grassi. Space Age Fertilizer: Landus Green Ammonia Plant Online in Iowa. thedailyscoop.com/news/retail-business/space-age-fertilizer-landus-green-ammonia-plant-online-iowa, 2024.

[114] Rachel Gregg, Joseph Beach, and Rick Bernheim. Starfire energy's rapid ramp modular ammonia plant development status and trajectory. ammoniaenergy.org/paper/starfire-energys-rapid-ramp-modular-ammonia-plant-development-status-and-trajectory/, 2021.

[115] T Grundt and K Christiansen. Hydrogen by water electrolysis as basis for small scale ammonia production. a comparison with hydrocarbon based technologies. *International Journal of Hydrogen Energy*, 7(3):247–257, 1982.

[116] Jianping Guo, Fei Chang, Peikun Wang, Daqiang Hu, Pei Yu, Guotao Wu, Zhitao Xiong, and Ping Chen. Highly active mnn–li2nh composite catalyst for producing co x-free hydrogen. *ACS Catalysis*, 5(5):2708–2713, 2015.

[117] Jianping Guo and Ping Chen. Interplay of alkali, transition metals, nitrogen, and hydrogen in ammonia synthesis and decomposition reactions. *Accounts of Chemical Research*, 54(10):2434–2444, 2021.

[118] Jianping Guo, Zheng Chen, Anan Wu, Fei Chang, Peikun Wang, Daqiang Hu, Guotao Wu, Zhitao Xiong, Pei Yu, and Ping Chen. Electronic promoter or reacting species? the role of linh 2 on ru in catalyzing nh 3 decomposition. *Chemical Communications*, 51(82):15161–15164, 2015.

[119] Jianping Guo, Peikun Wang, Guotao Wu, Anan Wu, Daqiang Hu, Zhitao Xiong, Junhu Wang, Pei Yu, Fei Chang, Zheng Chen, et al. Lithium imide synergy with 3d transition-metal nitrides leading to unprecedented catalytic activities for ammonia decomposition. *Angewandte Chemie*, 127(10):2993–2997, 2015.

[120] F Haber and R Le Rossignol. The production of synthetic ammonia. *Industrial and Engineering Chemistry*, 5(4):328–331, 1913.

[121] Ioannis Hadjipaschalis, Andreas Poullikkas, and Venizelos Efthimiou. Overview of current and future energy storage technologies for electric power applications. *Renewable and Sustainable Energy Reviews*, 13(6-7):1513–1522, 2009.

[122] R. Haghbakhsh, H. Peyrovedin, S. Raeissi, A. R. C. Duarte, and A. Shariati. Investigating the performance of novel green solvents in absorption refrigeration cycles: Energy and exergy analyses. *International Journal of Refrigeration*, 113:174–186, 2020.

[123] Wenfeng Han, Zhi Li, and Huazhang Liu. La2ce2o7 supported ruthenium as a robust catalyst for ammonia synthesis. *Journal of Rare Earths*, 37(5):492–499, 2019.

[124] Xue Han, Sihai Yang, and Martin Schroder. Metal–organic framework materials for production and distribution of ammonia. *Journal of the American Chemical Society*, 145(4):1998–2012, 2023.

[125] Ragnhild Hancke, Thomas Holm, and Øystein Ulleberg. The case for high-pressure pem water electrolysis. *Energy Conversion and Management*, 261:115642, 2022.

[126] Christoph Hank, André Sternberg, Nikolas Köppel, Marius Holst, Tom Smolinka, Achim Schaadt, Christopher Hebling, and Hans-Martin Henning. Energy efficiency and economic assessment of imported energy carriers based on renewable electricity. *Sustainable Energy & Fuels*, 4(5):2256–2273, 2020.

[127] Julia Hansson, Selma Brynolf, Erik Fridell, and Mariliis Lehtveer. The potential role of ammonia as marine fuel—based on energy systems modeling and multi-criteria decision analysis. *Sustainability*, 12(8):3265, 2020.

[128] Michikazu Hara, Masaaki Kitano, and Hideo Hosono. Ru-loaded c12a7: e^--electride as a catalyst for ammonia synthesis. *ACS Catalysis*, 7(4):2313–2324, 2017.

[129] Masashi Hattori, Taiyo Mori, Tomohiro Arai, Yasunori Inoue, Masato Sasase, Tomofumi Tada, Masaaki Kitano, Toshiharu Yokoyama, Michikazu Hara, and Hideo Hosono. Enhanced catalytic ammonia synthesis with transformed bao. *ACS Catalysis*, 8(12):10977–10984, 2018.

[130] Anne Hauch, Rainer Küngas, Peter Blennow, Anders Bavnhøj Hansen, John Bøgild Hansen, Brian Vad Mathiesen, and Mogens Bjerg Mogensen. Recent advances in solid oxide cell technology for electrolysis. *Science*, 370(6513), 2020.

[131] Fumitaka Hayashi, Masaaki Kitano, Toshiharu Yokoyama, Michikazu Hara, and Hideo Hosono. Surface treatment for conductive 12 cao·7 al2o3 electride powder by rapid thermal annealing processing and its application to ammonia synthesis. *ChemCatChem*, 6(5):1317–1323, 2014.

[132] Yuankang He and EL Cussler. Ammonia permeabilities of perfluorosulfonic membranes in various ionic forms. *Journal of Membrane Science*, 68(1-2):43–52, 1992.

[133] Jarkko Helminen, Joni Helenius, Erkki Paatero, and Ilkka Turunen. Comparison of sorbents and isotherm models for nh3-gas separation by adsorption. *AIChE Journal*, 46(8):1541–1555, 2000.

[134] Jarkko Helminen, Joni Helenius, Erkki Paatero, and Ilkka Turunen. Adsorption equilibria of ammonia gas on inorganic and organic sorbents at 298.15 k. *Journal of Chemical and Engineering Data*, 46(2):391–399, 2001.

[135] Haruo Hikita, Haruo Ishikawa, Takashi Okamoto, and Kiyotaka Mogami. Kinetics of simultaneous absorption of ethylene and chlorine into water and hydrochloric acid in an agitated vessel with a flat gas-liquid interface. *Industrial & Engineering Chemistry Research*, 26(9):1796–1803, 1987.

[136] Heath H Himstedt, Mark S Huberty, Alon V McCormick, Lanny D Schmidt, and EL Cussler. Ammonia synthesis enhanced by magnesium chloride absorption. *AIChE Journal*, 61(4):1364–1371, 2015.

[137] HNH Energy. Our Project. hnhenergy.com/en/, 2024.

[138] Tuan-Dung Hoang and Nhuan Nghiem. Recent developments and current status of commercial production of fuel ethanol. *Fermentation*, 7(4):314, 2021.

[139] Hideo Hosono, Michikazu Hara, Masaaki Kitano, Toshiharu Yokoyama, and Yasunori Inoue. Composite, method for producing composite, ammonia synthesis catalyst, and ammonia synthesis method, 2019. US Patent 10,322,940.

[140] Hideo Hosono, Michikazu Hara, Masaaki Kitano, Toshiharu Yokoyama, Yasunori Inoue, and Shinji Kambara. Supported metal catalyst and method of synthesizing ammonia using the same, 2019. US Patent 10,173,202.

[141] Hideo Hosono, Masaaki Kitano, Toshiharu Yokoyama, Jiang Li, Shigeki Kawamura, and Kazuhisa Kishida. Method for manufacturing ammonia synthesis catalyst, and method for manufacturing ammonia, 2024. US Patent 11,964,260.

[142] Daniel J Hrtus, Fouzia Hasan Nowrin, Austin Lomas, Yanick Fotsa, and Mahdi Malmali. Achieving+ 95 % ammonia purity by optimizing the absorption and desorption conditions of supported metal halides. *ACS Sustainable Chemistry & Engineering*, 10(1):204–212, 2021.

[143] Jia Huang, Jinjian Pan, Zhixiong You, and Xingmao Jiang. Enhanced ammonia synthesis activity of ru/ba5ta4o15 catalyst by reduction of ba5ta4o15 with cah2. *International Journal of Hydrogen Energy*, 47(65):28019–28024, 2022.

[144] Mark S Huberty, Andrew L Wagner, Alon McCormick, and Edward Cussler. Ammonia absorption at haber process conditions. *AIChE Journal*, 58(11):3526–3532, 2012.

[145] Jens S Hummelshøj, Rasmus Zink Sørensen, Marina Yu Kustova, Tue Johannessen, Jens K Nørskov, and Claus Hviid Christensen. Generation of nanopores during desorption of nh3 from mg (nh3) 6cl2. *Journal of the American Chemical Society*, 128(1):16–17, 2006.

[146] Stuart M Hunter, Duncan H Gregory, Justin SJ Hargreaves, Mélissandre Richard, Daniel Duprez, and Nicolas Bion. A study of 15n/14n isotopic exchange over cobalt molybdenum nitrides. *ACS Catalysis*, 3(8):1719–1725, 2013.

[147] Jussi Ikäheimo, Juha Kiviluoma, Robert Weiss, and Hannele Holttinen. Power-to-ammonia in future north european 100 % renewable power and heat system. *International Journal of Hydrogen Energy*, 43(36):17295–17308, 2018.

[148] Norihiko Iki, Osamu Kurata, Takayuki Matsunuma, Takahiro Inoue, Masato Suzuki, Taku Tsujimura, and Hirohide Furutani. Micro gas turbine firing kerosene and ammonia. In *Turbo Expo: Power for Land, Sea, and Air*, volume 56796, page V008T23A023. American Society of Mechanical Engineers, 2015.

[149] Mustafa Ilbas, AP Crayford, I Yılmaz, PJ Bowen, and N Syred. Laminar-burning velocities of hydrogen–air and hydrogen–methane–air mixtures: an experimental study. *International Journal of Hydrogen Energy*, 31(12):1768–1779, 2006.

[150] Kazuya Imamura, Shin-ichiro Miyahara, Yukiko Kawano, Katsutoshi Sato, Yuta Nakasaka, and Katsutoshi Nagaoka. Kinetics of ammonia synthesis over ru/pr2o3. *Journal of the Taiwan Institute of Chemical Engineers*, 105:50–56, 2019.

[151] Suheil Imtiaz. AM Green Ammonia Reaches Final Investment Decision (FID) on its first million-ton green ammonia project located in Kakinada, Andhra Pradesh, India. newswire.ca/news-releases/am-green-ammonia-reaches-final-investment-decision-fid-on-its-first-million-ton-green-ammonia-project-located-in-kakinada-andhra-pradesh-india-898888983.html, 2024.

[152] Yasunori Inoue, Masaaki Kitano, Sung-Wng Kim, Toshiharu Yokoyama, Michikazu Hara, and Hideo Hosono. Highly dispersed ru on electride [ca24al28o64] 4+(e–) 4 as a catalyst for ammonia synthesis. *ACS Catalysis*, 4(2):674–680, 2014.

[153] Yasunori Inoue, Masaaki Kitano, Kazuhisa Kishida, Hitoshi Abe, Yasuhiro Niwa, Masato Sasase, Yusuke Fujita, Hiroki Ishikawa, Toshiharu Yokoyama, Michikazu Hara, et al. Efficient and stable ammonia synthesis by self-organized flat ru nanoparticles on calcium amide. *ACS Catalysis*, 6(11):7577–7584, 2016.

[154] International Energy Agency. Emissions measurement and data collection for a net zero steel industry. iea.org/reports/emissions-measurement-and-data-collection-for-a-net-zero-steel-industry, 2023.

[155] International Energy Agency. Carbon Capture, Utilisation, and Storage. iea.org/energy-system/carbon-capture-utilisation-and-storage, 2024.

[156] International Energy Association. Electrolysers: Technology deep dive. iea.org/reports/electrolysers, 2022.

[157] International Energy Association. Hydrogen - Net Zero Emissions Guide. iea.org/reports/hydrogen-2156, 2023.

[158] International Energy Association. Solar PV. iea.org/energy-system/renewables/solar-pv, 2023.

[159] International Energy Association. Wind. iea.org/energy-system/renewables/wind, 2023.

[160] International Energy Association. Electrolysers. iea.org/energy-system/low-emission-fuels/electrolysers, 2024.

[161] International Energy Association. Hydroelectricity. iea.org/energy-system/renewables/hydroelectricity, 2024.

[162] International Energy Association. Welcome to Haldor topsoe's Vej 2, Home of the EU's largest industrial-scale SOEC manufacturing facility. topsoe.com/herning, 2025.

[163] KN Iost, VL Temerev, NS Smirnova, DA Shlyapin, VA Borisov, IV Muromtsev, MV Trenikhin, TV Kireeva, AV Shilova, and PG Tsyrul'nikov. Synthesis and study of ru–ba–cs/sibunit ternary catalysts for ammonia synthesis. *Russian Journal of Applied Chemistry*, 90:887–894, 2017.

[164] Svetlana Ivanova and Robert Lewis. Producing nitrogen via pressure swing adsorption. *Chemical Engineering Progress*, 108(6):38–42, 2012.

[165] Jun Iwamoto, Masahiro Itoh, Yoshio Kajita, Makoto Saito, and Ken-ichi Machida. Ammonia synthesis on magnesia supported ruthenium catalysts with mesoporous structure. *Catalysis Communications*, 8(6):941–944, 2007.

[166] Claus JH Jacobsen. Novel class of ammonia synthesis catalysts. *Chemical Communications*, (12):1057–1058, 2000.

[167] Claus JH Jacobsen. Boron nitride: a novel support for ruthenium-based ammonia synthesis catalysts. *Journal of Catalysis*, 200(1):1–3, 2001.

[168] Claus JH Jacobsen, Søren Dahl, Bjerne S Clausen, Sune Bahn, Ashildur Logadottir, and Jens K Nørskov. Catalyst design by interpolation in the periodic table: bimetallic ammonia synthesis catalysts. *Journal of the American Chemical Society*, 123(34):8404–8405, 2001.

[169] Hjalte Sylvest Jacobsen, Heine Anton Hansen, Jens Wenzel Andreasen, Qing Shi, Anders Andreasen, Robert Feidenhans, Martin Meedom Nielsen, Kenny Ståhl, and Tejs Vegge. Nanoscale structural characterization of mg (nh3) 6cl2 during nh3 desorption: An in situ small angle x-ray scattering study. *Chemical Physics Letters*, 441(4-6):255–260, 2007.

[170] P. Jaramillo, S. Kahn Ribeiro, P. Newman, S. Dhar, O. E. Diemuodeke, T. Kajino, D. S. Lee, S. B. Nugroho, X. Ou, A. Hammer Strømman, and J. Whitehead. *Climate Change 2022: Mitigation of Climate Change. Contribution of Working Group III to the Sixth Assessment Report of the Intergovernmental Panel on Climate Change*, chapter Transport. Cambridge University Press (Cambridge, UK and New York (NY), USA), 2022.

[171] J Richard Jennings. *Catalytic ammonia synthesis: Fundamentals and practice*. Springer Science & Business Media, 1991.

[172] Peter Bjerre Jensen, Steen Lysgaard, Ulrich J Quaade, and Tejs Vegge. Designing mixed metal halide ammines for ammonia storage using density functional theory and genetic algorithms. *Physical Chemistry Chemical Physics*, 16(36):19732–19740, 2014.

[173] John Cockerill. AM Green entrusts John Cockerill with India's largest electrolyzer order for its green ammonia complex, one of the world's largest. johncockerill.com/en/press-and-news/news/am-green-entrusts-john-cockerill-with-indias-largest-electrolyzer-order-for-its-green-ammonia-complex-one-of-the-worlds-largest, 2024.

[174] Rune E Johnsen, Peter B Jensen, Poul Norby, and Tejs Vegge. Temperature-and pressure-induced changes in the crystal structure of sr (nh3) 8cl2. *The Journal of Physical Chemistry C*, 118(42):24349–24356, 2014.

[175] Hiroshi Kageyama, Yoji Kobayashi, Naoya Masuda, and Hideo Hosono. Ammonia synthesis catalyst and ammonia synthesis method, July 10 2018. US Patent 10,017,395.

[176] Mototaka Kai, Yasushi Fujimuri, Takayoshi Fujimoto, Hideyuki Takagi, Yuichi Manaka, and Tetsuya Nanba. Demonstration of CO_2-Free Ammonia Synthesis Using Renewable Energy-Generated Hydrogen. nh3fuelassociation.org/2018/12/08/demonstration-of-co2-free-ammonia-synthesis-using-renewable-energy-generated-hydrogen/, 2018.

[177] Matthew J Kale, Deepak K Ojha, Sayandeep Biswas, Joshua I Militti, Alon V McCormick, Jeffrey H Schott, Paul J Dauenhauer, and EL Cussler. Optimizing ammonia separation via reactive absorption for sustainable ammonia synthesis. *ACS Applied Energy Materials*, 3(3):2576–2584, 2020.

[178] Yildiz Kalinci, Arif Hepbasli, and Ibrahim Dincer. Biomass-based hydrogen production: a review and analysis. *International Journal of Hydrogen Energy*, 34(21):8799–8817, 2009.

[179] Seamus P Kane, Darrick Zarling, and William F Northrop. Thermochemical and sensible energy recuperation using thermally-integrated reactor and diesel-ammonia dual fueling strategy. *Internal Combustion Engine Division Fall Technical Conference Proceedings*, 59346:V001T02A011, 2019.

[180] Masakoto Kanezashi, Akira Yamamoto, Tomohisa Yoshioka, and Toshinori Tsuru. Characteristics of ammonia permeation through porous silica membranes. *AIChE Journal*, 56(5):1204–1212, 2010.

[181] ML Kastens and WG McBurney. Calcium cyanamide. *Industrial and Engineering Chemistry*, 43(5):1020–1033, 1951.

[182] Michael J Katz, Ashlee J Howarth, Peyman Z Moghadam, Jared B DeCoste, Randall Q Snurr, Joseph T Hupp, and Omar K Farha. High volumetric uptake of ammonia using cu-mof-74/cu-cpo-27. *Dalton Transactions*, 45(10):4150–4153, 2016.

[183] Martin Keller, Mitsuo Koshi, Junichiro Otomo, Hiroshi Iwasaki, Teruo Mitsumori, and Koichi Yamada. Thermodynamic evaluation of an ammonia-fueled combined-cycle gas turbine process operated under fuel-rich conditions. *Energy*, 194:116894, 2020.

[184] Mohammad Hasan Khademi and Reyhaneh Sadat Sabbaghi. Comparison between three types of ammonia synthesis reactor configurations in terms of cooling methods. *Chemical Engineering Research & Design*, 128:306–317, 2017.

[185] Faisal I Khan and Aloke Kr Ghoshal. Removal of volatile organic compounds from polluted air. *Journal of Loss Prevention in the Process Industries*, 13(6):527–545, 2000.

[186] John A Kilner, Stephen J Skinner, SJC Irvine, and PP Edwards. *Functional materials for sustainable energy applications*. Elsevier, 2012.

[187] Dae Won Kim, Dong Won Kang, Minjung Kang, Jung-Hoon Lee, Jong Hyeak Choe, Yun Seok Chae, Doo San Choi, Hongryeol Yun, and Chang Seop Hong. High ammonia uptake of a metal–organic framework adsorbent in a wide pressure range. *Angewandte Chemie International Edition*, 59(50):22531–22536, 2020.

[188] Kazuhisa Kishida, Masaaki Kitano, Masato Sasase, Peter V Sushko, Hitoshi Abe, Yasuhiro Niwa, Kiya Ogasawara, Toshiharu Yokoyama, and Hideo Hosono. Air-stable calcium cyanamide-supported ruthenium catalyst for ammonia synthesis and decomposition. *ACS Applied Energy Materials*, 3(7):6573–6582, 2020.

[189] Masaaki Kitano, Yasunori Inoue, Hiroki Ishikawa, Kyosuke Yamagata, Takuya Nakao, Tomofumi Tada, Satoru Matsuishi, Toshiharu Yokoyama, Michikazu Hara, and Hideo Hosono. Essential role of hydride ion in ruthenium-based ammonia synthesis catalysts. *Chemical Science*, 7(7):4036–4043, 2016.

[190] Masaaki Kitano, Yasunori Inoue, Youhei Yamazaki, Fumitaka Hayashi, Shinji Kanbara, Satoru Matsuishi, Toshiharu Yokoyama, Sung-Wng Kim, Michikazu Hara, and Hideo Hosono. Ammonia synthesis using a stable electride as an electron donor and reversible hydrogen store. *Nature Chemistry*, 4(11):934–940, 2012.

[191] Masaaki Kitano, Shinji Kanbara, Yasunori Inoue, Navaratnarajah Kuganathan, Peter V Sushko, Toshiharu Yokoyama, Michikazu Hara, and Hideo Hosono. Electride support boosts nitrogen dissociation over ruthenium catalyst and shifts the bottleneck in ammonia synthesis. *Nature Communications*, 6(1):6731, 2015.

[192] Masaaki Kitano, Jun Kujirai, Kiya Ogasawara, Satoru Matsuishi, Tomofumi Tada, Hitoshi Abe, Yasuhiro Niwa, and Hideo Hosono. Low-temperature synthesis of perovskite oxynitride-hydrides as ammonia synthesis catalysts. *Journal of the American Chemical Society*, 141(51):20344–20353, 2019.

[193] Asbjørn Klerke, Claus Hviid Christensen, Jens K Nørskov, and Tejs Vegge. Ammonia for hydrogen storage: challenges and opportunities. *Journal of Materials Chemistry*, 18(20):2304–2310, 2008.

[194] Hideaki Kobayashi, Akihiro Hayakawa, KD Kunkuma A Somarathne, and Ekenechukwu C Okafor. Science and technology of ammonia combustion. *Proceedings of the Combustion Institute*, 37(1):109–133, 2019.

[195] Olesia Kozachuk, Ignacio Luz, Francesc X Llabrés i Xamena, Heshmat Noei, Max Kauer, H Bauke Albada, Eric D Bloch, Bernd Marler, Yuemin Wang, Martin Muhler, et al. Multifunctional, defect-engineered metal–organic frameworks with ruthenium centers: sorption and catalytic properties. *Angewandte Chemie International Edition*, 53(27):7058–7062, 2014.

[196] Olesia Kozachuk, Kirill Yusenko, Heshmat Noei, Yuemin Wang, Stephan Walleck, Thorsten Glaser, and Roland A Fischer. Solvothermal growth of a ruthenium metal–organic framework featuring hkust-1 structure type as thin films on oxide surfaces. *Chemical Communications*, 47(30):8509–8511, 2011.

[197] R Krabetz and Cl Peters. Zur funktion der aktivatoren im technischen ammoniak-katalysator. *Angewandte Chemie*, 77(7):333–339, 1965.

[198] Osamu Kurata, Norihiko Iki, Takayuki Matsunuma, Takahiro Inoue, Taku Tsujimura, Hirohide Furutani, Hideaki Kobayashi, and Akihiro Hayakawa. Performances and emission characteristics of NH_3–air and NH_3CH_4–air combustion gas-turbine power generations. *Proceedings of the Combustion Institute*, 36(3):3351–3359, 2017.

[199] Daniel V Laciak, Guido P Pez, and Peter M Burban. Molten salt facilitated transport membranes. part 2. separation of ammonia from nitrogen and hydrogen at high temperatures. *Journal of Membrane Science*, 65(1-2):31–38, 1992.

[200] MA Laguna-Bercero. Recent advances in high temperature electrolysis using solid oxide fuel cells: A review. *Journal of Power Sources*, 203:4–16, 2012.

[201] M Laimon and S Goh. Unlocking potential in renewable energy curtailment for green ammonia production. *International Journal of Hydrogen Energy*, 71:964–971, 2024.

[202] Krystina E Lamb, Michael D Dolan, and Danielle F Kennedy. Ammonia for hydrogen storage; A review of catalytic ammonia decomposition and hydrogen separation and purification. *International Journal of Hydrogen Energy*, 44(7):3580–3593, 2019.

[203] Penghang Lan, She Chen, Qihang Li, Kelin Li, Feng Wang, Yaoxun Zhao, and Tianwei Wang. Comparison of different hydrogen-ammonia energy conversion pathways for renewable energy supply. *Renewable Energy*, 227:120602, 2024.

[204] Rong Lan, John TS Irvine, and Shanwen Tao. Ammonia and related chemicals as potential indirect hydrogen storage materials. *International Journal of Hydrogen Energy*, 37(2):1482–1494, 2012.

[205] Rong Lan and Shanwen Tao. Ammonia as a suitable fuel for fuel cells. *Frontiers in Energy Research*, 2:35, 2014.

[206] M Soledad Larrechi, Andry Cera-Manjarres, and Alberto Coronas. Ranking the solubility of ammonia in ionic liquids using near infrared spectroscopy and multivariate curve resolution. *Spectrochimica Acta. Part A: Molecular and Biomolecular Spectroscopy*, 215:88–96, 2019.

[207] Lazard. 2023 Levelized Cost Of Energy+. lazard.com/research-insights/2023-levelized-cost-of-energyplus/, 2023.

[208] Lazard. Levelized Cost of Energy+. lazard.com/research-insights/levelized-cost-of-energyplus/, 2024.

[209] Markus Lehner, Robert Tichler, Horst Steinmüller, and Markus Koppe. *Power-to-gas: technology and business models*. Springer, 2014.

[210] HY Leng, Takayuki Ichikawa, Satoshi Hino, Nobuko Hanada, Shigehito Isobe, and Hironobu Fujii. Synthesis and decomposition reactions of metal amides in metal–n–h hydrogen storage system. *Journal of Power Sources*, 156(2):166–170, 2006.

[211] Chunyan Li, Zecheng Zhang, Yiping Zheng, Biyun Fang, Jun Ni, Jianxin Lin, Bingyu Lin, Xiuyun Wang, and Lilong Jiang. Titanium modified ru/ceo2 catalysts for ammonia synthesis. *Chemical Engineering Science*, 251:117434, 2022.

[212] Jiang Li, Jiazhen Wu, Haiyun Wang, Yangfan Lu, Tiannan Ye, Masato Sasase, Xiaojun Wu, Masaaki Kitano, Takeshi Inoshita, and Hideo Hosono. Acid-durable electride with layered ruthenium for ammonia synthesis: boosting the activity via selective etching. *Chemical Science*, 10(22):5712–5718, 2019.

[213] Jinpeng Li, Weiyang Wang, Wenxing Chen, Qinmei Gong, Jun Luo, Ruoqian Lin, Huolin Xin, Hui Zhang, Dingsheng Wang, Qing Peng, et al. Sub-nm ruthenium cluster as an efficient and robust catalyst for decomposition and synthesis of ammonia: Break the "size shackles". *Nano Research*, 11:4774–4785, 2018.

[214] Wei Li, Pengcheng Liu, Ruyue Niu, Jinping Li, and Shuang Wang. Influence of ceo2 supports prepared with different precipitants over ru/ceo2 catalysts for ammonia synthesis. *Solid State Sciences*, 99:105983, 2020.

[215] Wei Li, Shuang Wang, and Jinping Li. Highly effective ru/baceo3 catalysts on supports with strong basic sites for ammonia synthesis. *Chemistry–An Asian Journal*, 14(16):2815–2821, 2019.

[216] Zhijie Li, Xiangping Zhang, Haifeng Dong, Xiaochun Zhang, Hongshuai Gao, Suojiang Zhang, Jianwei Li, and Congmin Wang. Efficient absorption of ammonia with hydroxyl-functionalized ionic liquids. *RSC Advances*, 5(99):81362–81370, 2015.

[217] Zi-Liang Li, Fu-Yu Zhong, Lin-Sen Zhou, Zi-Qi Tian, and Kuan Huang. Deep eutectic solvents formed by n-methylacetamide and heterocyclic weak acids for highly efficient and reversible chemical absorption of ammonia. *Industrial & Engineering Chemistry Research*, 59(5):2060–2067, 2020.

[218] Bingyu Lin, Yunjie Guo, Chenfeng Cao, Jun Ni, Jianxin Lin, and Lilong Jiang. Carbon support surface effects in the catalytic performance of ba-promoted ru catalyst for ammonia synthesis. *Catalysis Today*, 316:230–236, 2018.

[219] Bingyu Lin, Lan Heng, Biyun Fang, Haiyun Yin, Jun Ni, Xiuyun Wang, Jianxin Lin, and Lilong Jiang. Ammonia synthesis activity of alumina-supported ruthenium catalyst enhanced by alumina phase transformation. *ACS Catalysis*, 9(3):1635–1644, 2019.

[220] Bingyu Lin, Lan Heng, Haiyun Yin, Biyun Fang, Jun Ni, Xiuyun Wang, Jianxin Lin, and Lilong Jiang. Effects of using carbon-coated alumina as support for ba-promoted ru catalyst in ammonia synthesis. *Industrial & Engineering Chemistry Research*, 58(24):10285–10295, 2019.

[221] Bingyu Lin, Rong Wang, Jianxin Lin, Jun Ni, and Kemei Wei. Effect of carbon and chlorine on the performance of carbon-covered alumina supported ru catalyst for ammonia synthesis. *Catalysis Communications*, 12(15):1452–1457, 2011.

[222] Bosong Lin, Fouzia Hasan Nowrin, Justin J Rosenthal, Abhoyjit S Bhown, and Mahdi Malmali. Perspective on intensification of haber- bosch to enable ammonia production under milder conditions. *ACS Sustainable Chemistry & Engineering*, 11(27):9880–9899, 2023.

[223] Bosong Lin, Theodore Wiesner, and Mahdi Malmali. Performance of a small-scale haber process: A techno-economic analysis. *ACS Sustainable Chemistry & Engineering*, 8(41):15517–15531, 2020.

[224] J Lin, Ziqing Wang, Liuming Zhang, Ni Jun, Wang Rong, and Wei Kemei. Ammonia synthesis over ruthenium catalysts using barium-doped zirconia as supports prepared by citric acid method. *Chinese Journal of Catalysis*, 33(7-8):1075–1079, 2012.

[225] Chun Yi Liu and Ken-ichi Aika. Absorption and desorption behavior of ammonia with alkali earth halide and mixed halide. *Chemistry Letters*, 31(8):798–799, 2002.

[226] Chun Yi Liu and Ken-ichi Aika. Ammonia adsorption on ion exchanged y-zeolites as ammonia storage material. *Journal of the Japan Petroleum Institute*, 46(5):301–307, 2003.

[227] Chun Yi Liu and Ken-ichi Aika. Ammonia absorption into alkaline earth metal halide mixtures as an ammonia storage material. *Industrial & Engineering Chemistry Research*, 43(23):7484–7491, 2004.

[228] Chun Yi Liu and Ken-ichi Aika. Ammonia absorption on alkaline earth halides as ammonia separation and storage procedure. *Bulletin of the Chemical Society of Japan*, 77(1):123–131, 2004.

[229] Lin Liu, Xilun Zhang, Xiaohua Ju, Ji Feng, Jiemin Wang, and Ping Chen. Ru nanoparticles on a cs-loaded mgo superbase as highly efficient catalysts for ammonia synthesis. *Dalton Transactions*, 50(35):12074–12078, 2021.

[230] Pengcheng Liu, Ruyue Niu, Wei Li, Shuang Wang, and Jinping Li. The effect of barium-promoted for microsphere ru/ceo2 catalysts in ammonia synthesis. *Energy Sources, Part A: Recovery, Utilization, and Environmental Effects*, 41(6):689–699, 2019.

[231] Pengcheng Liu, Ruyue Niu, Wei Li, Shuang Wang, and Jinping Li. Morphology effect of ceria on the ammonia synthesis activity of ru/ceo 2 catalysts. *Catalysis Letters*, 149:1007–1016, 2019.

[232] Xinyu Liu, Amgad Elgowainy, and Michael Wang. Life cycle energy use and greenhouse gas emissions of ammonia production from renewable resources and industrial by-products. *Green Chemistry*, 22(17):5751–5761, 2020.

[233] Yangfan Lu, Jiang Li, Tomofumi Tada, Yoshitake Toda, Shigenori Ueda, Toshiharu Yokoyama, Masaaki Kitano, and Hideo Hosono. Water durable electride y5si3: electronic structure and catalytic activity for ammonia synthesis. *Journal of the American Chemical Society*, 138(12):3970–3973, 2016.

[234] Jolie M Lucero, James M Crawford, Colin A Wolden, and Moises A Carreon. Tunability of ammonia adsorption over nap zeolite. *Microporous and Mesoporous Materials*, 324:111288, 2021.

[235] Li Luo, Jiaran Li, Xiaoyan Chen, Xuegong Cao, Yibang Liu, Zhixin Wu, Xiaoyan Luo, and Congmin Wang. Superhigh and reversible nh3 uptake of cobaltous thiocyanate functionalized porous poly ionic liquids through competitive and cooperative interactions. *Chemical Engineering Journal*, 427:131638, 2022.

[236] Li Luo, Zhihong Wu, Zhixin Wu, Yibang Liu, Xiaoyun Huang, Renhui Ling, Ling Ye, Xiaoyan Luo, and Congmin Wang. Role of structure in the ammonia uptake of porous polyionic liquids. *ACS Sustainable Chemistry & Engineering*, 10(13):4094–4104, 2022.

[237] Qun Luo, Jianding Li, Bo Li, Bin Liu, Huaiyu Shao, and Qian Li. Kinetics in mg-based hydrogen storage materials: Enhancement and mechanism. *Journal of Magnesium and Alloys*, 7(1):58–71, 2019.

[238] Ignacio Luz, Sameer Parvathikar, Michael Carpenter, Timothy Bellamy, Kelly Amato, John Carpenter, and Marty Lail. Mof-derived nanostructured catalysts for low-temperature ammonia synthesis. *Catalysis Science & Technology*, 10(1):105–112, 2020.

[239] Yongcheng Ma, Guojun Lan, Xiaolong Wang, Geshan Zhang, Wenfeng Han, Haodong Tang, Huazhang Liu, and Ying Li. Effect of nitrogen co-doping with ruthenium on the catalytic performance of ba/ru-n-mc catalysts for ammonia synthesis. *RSC Advances*, 9(38):22045–22052, 2019.

[240] Yongde Ma, Jiayin Zhang, Kuan Huang, and Lilong Jiang. Highly efficient and selective separation of ammonia by deep eutectic solvents through cooperative acid-base and strong hydrogen-bond interaction. *Journal of Molecular Liquids*, 337:116463, 2021.

[241] Yujie Ma, Wanpeng Lu, Xue Han, Yinlin Chen, Ivan Da Silva, Daniel Lee, Alena M Sheveleva, Zi Wang, Jiangnan Li, Weiyao Li, et al. Direct observation of ammonia storage in uio-66 incorporating cu (ii) binding sites. *Journal of the American Chemical Society*, 144(19):8624–8632, 2022.

[242] Zhanwei Ma, Shengli Zhao, Xiaoping Pei, Xumao Xiong, and Bin Hu. New insights into the support morphology-dependent ammonia synthesis activity of ru/ceo 2 catalysts. *Catalysis Science & Technology*, 7(1):191–199, 2017.

[243] SM Tanveer Mahtab and Mahdi Malmali. Insights into structure-composition-capacity relationship of ultrastable metal halide absorbents for ammonia separation and storage. *Chemical Engineering Journal*, page 160105, 2025.

[244] Camel Makhloufi and Nouaamane Kezibri. Large-scale decomposition of green ammonia for pure hydrogen production. *International Journal of Hydrogen Energy*, 46(70):34777–34787, 2021.

[245] Mahdi Malmali, Giang Le, Jennifer Hendrickson, Joshua Prince, Alon V McCormick, and Edward L Cussler. Better absorbents for ammonia separation. *ACS Sustainable Chemistry & Engineering*, 6(5):6536–6546, 2018.

[246] Mahdi Malmali, Michael Reese, Alon V McCormick, and EL Cussler. Converting wind energy to ammonia at lower pressure. *ACS Sustainable Chemistry & Engineering*, 6(1):827–834, 2018.

[247] Mahdi Malmali, Yongming Wei, Alon McCormick, and Edward L Cussler. Ammonia synthesis at reduced pressure via reactive separation. *Industrial & Engineering Chemistry Research*, 55(33):8922–8932, 2016.

[248] Mantec Consultants PVT LTD. Installation of GTG cum HRSG under Energy Reduction Scheme at Nangal Unit. environmentclearance.nic.in/writereaddata/Online/TOR/30_Dec_2017_111158050Q9EY8ZWNPFR-NFLNangalGTG_HRSG.pdf, 2017.

[249] Mariano Martín and Antonio Sánchez. Biomass pathways to produce green ammonia and urea. *Current Opinion in Green and Sustainable Chemistry*, page 100933, 2024.

[250] Katie McDonald and Eric Guter. Air Products, ACWA Power and NEOM Sign Agreement for $5 Billion Production Facility in NEOM Powered by Renewable Energy for Production and Export of Green Hydrogen to Global Markets. airproducts.com/company/news-center/2020/07/0707-air-products-agreement-for-green-ammonia-production-facility-for-export-to-hydrogen-market, 2020.

[251] Andrew J Medford, Aleksandra Vojvodic, Jens S Hummelshøj, Johannes Voss, Frank Abild-Pedersen, Felix Studt, Thomas Bligaard, Anders Nilsson, and Jens K Nørskov. From the sabatier principle to a predictive theory of transition-metal heterogeneous catalysis. *Journal of Catalysis*, 328:36–42, 2015.

[252] Mejillones Ammonia Energy. Project Volta. https://mae-energy.com/en/proyecto-volta/, 2024.

[253] Abdelrahman Mewafy, Islam Ismael, Sahar S Kaddah, Weihao Hu, Zhe Chen, and Sayed Abulanwar. Optimal design of multiuse hybrid microgrids power by green hydrogen–ammonia. *Renewable and Sustainable Energy Reviews*, 192:114174, 2024.

[254] Jason Miller, William L Luyben, Paul Belanger, Stephane Blouin, and Larry Megan. Improving agility of cryogenic air separation plants. *Industrial & Engineering Chemistry Research*, 47(2):394–404, 2008.

[255] Stefano Mingolla, Kevin Rouwenhorst, Paolo Gabrielli, Giovanni Sansavini, Magdalena M Klemun, and Zhongming Lu. Towards sustainable fertilizer production: Cost comparison of flexible and continuous electrolytic ammonia production. In *15th International Conference on Applied Energy, ICAE 2023*. Energy Proceedings, 2024.

[256] Ignacio Luz Minguez, Mustapha Soukri, Marty Lail, John R Carpenter, Sameer Parvathikar, and Michael Carpenter. Method of making confined nanocatalysts within mesoporous materials and uses thereof, 2022. US Patent 11,406,971.

[257] Mariagiovanna Minutillo, Alessandra Perna, and Alessandro Sorce. Green hydrogen production plants via biogas steam and autothermal reforming processes: energy and exergy analyses. *Applied Energy*, 277:115452, 2020.

[258] Alwin Mittasch and W Frankenburg. Early studies of multicomponent catalysts. In *Advances in catalysis*, volume 2, pages 81–104. Elsevier, 1950.

[259] Hiroshi Mizoguchi, Masaaki Okunaka, Masaaki Kitano, Satoru Matsuishi, Toshiharu Yokoyama, and Hideo Hosono. Hydride-based electride material, lnh2 (ln= la, ce, or y). *Inorganic Chemistry*, 55(17):8833–8838, 2016.

[260] Christian Sandersen Mørch, Andreas Bjerre, Morten Piil Gøttrup, Spencer C Sorenson, and Jesper Schramm. Ammonia/hydrogen mixtures in an si-engine: Engine performance and analysis of a proposed fuel system. *Fuel*, 90(2):854–864, 2011.

[261] Eric R. Morgan. *Techno-economic feasibility study of ammonia plants powered by offshore wind*. PhD thesis, University of Massachusetts Amherst, 2013.

[262] Natalia Morlanés, Walid Almaksoud, Rohit K Rai, Samy Ould-Chikh, Mohammed M Ali, Balamurugan Vidjayacoumar, Bedour E Al-Sabban, Khalid Albahily, and Jean-Marie Basset. Development of catalysts for ammonia synthesis based on metal phthalocyanine materials. *Catalysis Science & Technology*, 10(3):844–852, 2020.

[263] Maria T Mota-Martinez, Patrick Brandl, Jason P Hallett, and Niall Mac Dowell. Challenges and opportunities for the utilisation of ionic liquids as solvents for co 2 capture. *Molecular Systems Design & Engineering*, 3(3):560–571, 2018.

[264] Theodore Motyka (Savannah River National Laboratory). Metal Hydrides. energy.gov/sites/default/files/2015/02/f19/fcto_h2_storage_summit_motyka.pdf, 2015.

[265] Martin Muhler, Frank Rosowski, Olaf Hinrichsen, Alessandra Hornung, and Gerhard Ertl. Ruthenium as catalyst for ammonia synthesis. In *Studies in Surface Science and Catalysis*, volume 101, pages 317–326. Elsevier, 1996.

[266] Tonje Naess and Hilde Steinfeld. Yara opens renewable hydrogen plant: "A major milestone". yara.com/corporate-releases/yara-opens-renewable-hydrogen-plant-a-major-milestone/, 2024.

[267] Katabathini Narasimharao, Podila Seetharamulu, KS Rama Rao, and Sulaiman N Basahel. Carbon covered mg–al hydrotalcite supported nanosized ru catalysts for ammonia synthesis. *Journal of Molecular Catalysis. A, Chemical*, 411:157–166, 2016.

[268] National Renewable Energy Laboratory. Commercial and Residential Hourly Load Profiles for all TMY3 Locations in the United States. doi.org/10.25984/1788456, 2014.

[269] National Renewable Energy Laboratory. Annual Technology Baseline. atb.nrel.gov/electricity/2022/about, 2023.

[270] National Renewable Energy Laboratory. Energy Analysis: Utility-Scale Energy Technology Capacity Factors. nrel.gov/analysis/tech-cap-factor.html, 2023.

[271] National Renewable Energy Laboratory. Typical Meteorological Year. nsrdb.nrel.gov/data-sets/tmy, 2023.

[272] Richard Nayak-Luke, René Bañares-Alcántara, and Ian Wilkinson. "green" ammonia: Impact of renewable energy intermittency on plant sizing and levelized cost of ammonia. *Industrial & Engineering Chemistry Research*, 57(43):14607–14616, 2018.

[273] Richard M Nayak-Luke and René Bañares-Alcántara. Long-term energy storage: What is the need and is ammonia a solution? In *Computer Aided Chemical Engineering*, volume 44, pages 1843–1848. Elsevier, 2018.

[274] Richard Michael Nayak-Luke and René Bañares-Alcántara. Techno-economic viability of islanded green ammonia as a carbon-free energy vector and as a substitute for conventional production. *Energy & Environmental Science*, 13(9):2957–2966, 2020.

[275] NEOM. NEOM Green Hydrogen Company completes financial close at a total investment value of USD 8.4 billion in the world's largest carbon-free green hydrogen plant. neom.com/en-us/newsroom/neom-green-hydrogen-investment, 2023.

[276] Jun Ni, Baoqiang Jing, Jianxin Lin, Bingyu Lin, Ziqi Zhao, and Lilong Jiang. Effect of rare earth on the performance of ru/mgal-ldo catalysts for ammonia synthesis. *Journal of Rare Earths*, 36(2):135–141, 2018.

[277] Jun Ni, Siyu Shi, Chuanfeng Zhang, Biyun Fang, Xiuyun Wang, Jianxin Lin, Shijing Liang, Bingyu Lin, and Lilong Jiang. Enhanced catalytic performance of the carbon-supported ru ammonia synthesis catalyst by an introduction of oxygen functional groups via gas-phase oxidation. *Journal of Catalysis*, 409:78–86, 2022.

[278] Masayasu Nishi, Shih-Yuan Chen, and Hideyuki Takagi. Energy efficient and intermittently variable ammonia synthesis over mesoporous carbon-supported cs-ru nanocatalysts. *Catalysts*, 9(5):406, 2019.

[279] Masayasu Nishi, Shih-Yuan Chen, and Hideyuki Takagi. Mild ammonia synthesis over ba-promoted ru/mpc catalysts: Effects of the ba/ru ratio and the mesoporous structure. *Catalysts*, 9(5):480, 2019.

[280] Masayasu Nishi, Shih-Yuan Chen, Hiroyuki Tateno, Takehisa Mochizuki, Hideyuki Takagi, and Tetsuya Nanba. A super-growth carbon nanotubes-supported, cs-promoted ru catalyst for 0.1–8 mpag ammonia synthesis. *Journal of Catalysis*, 413:623–635, 2022.

[281] Tejas Nivarty, William C Straub, Chinomso E Onuoha, Michael Manno, Jeffrey H Schott, Mahdi Malmali, and Alon V McCormick. Advancing sustainable ammonia production via solventless, robust, and thermally conductive absorbents. *ACS Sustainable Chemistry & Engineering*, 13(13):4955–4962, 2025.

[282] Fouzia Hasan Nowrin and Mahdi Malmali. Optimizing reaction-absorption process for lower pressure ammonia production. *ACS Sustainable Chemistry & Engineering*, 10(37):12319–12328, 2022.

[283] Yuta Ogura, Katsutoshi Sato, Shin-ichiro Miyahara, Yukiko Kawano, Takaaki Toriyama, Tomokazu Yamamoto, Syo Matsumura, Saburo Hosokawa, and Katsutoshi Nagaoka. Efficient ammonia synthesis over a ru/la 0.5 ce 0.5 o 1.75 catalyst pre-reduced at high temperature. *Chemical Science*, 9(8):2230–2237, 2018.

[284] Deepak K Ojha, Matthew J Kale, Alon V McCormick, Michael Reese, Mahdi Malmali, Paul Dauenhauer, and EL Cussler. Integrated ammonia synthesis and separation. *ACS Sustainable Chemistry & Engineering*, 7(23):18785–18792, 2019.

[285] Chinomso E Onuoha, Matthew J Kale, Mahdi Malmali, Paul J Dauenhauer, and Alon V McCormick. Improving absorbent-enhanced ammonia separation for efficient small-scale ammonia synthesis. *Industrial & Engineering Chemistry Research*, 63(13):5608–5617, 2024.

[286] Chinomso E Onuoha, William C Straub, Tejas Nivarty, Mahdi Malmali, Paul J Dauenhauer, and Alon V McCormick. Enhancing green ammonia capture and storage: Structure and composition of thermally conductive mgcl2–al absorbents for reproducible cyclic performance. *ACS Applied Energy Materials*, 8(7):4103–4112, 2025.

[287] Ola Osman, Sgouris Sgouridis, and Andrei Sleptchenko. Scaling the production of renewable ammonia: A techno-economic optimization applied in regions with high insolation. *Journal of Cleaner Production*, 271:121627, 2020.

[288] Joe L Outlaw, Bart L Fischer, Henry L Bryant, and Marc J Raulston. Economic Impact of Nitrogen Prices on U.S. Corn Producers. dt176nijwh14e.cloudfront.net/file/481/Study%20.pdf, 2022.

[289] Wei Ouyang, Songsheng Zheng, Chongjun Wu, Xiaohui Hu, Riyi Chen, Lianghui Zhuo, and Zhaolin Wang. Dynamic ammonia adsorption by fau zeolites to below 0.1 ppm for hydrogen energy applications. *International Journal of Hydrogen Energy*, 46(64):32559–32569, 2021.

[290] Ahmet Ozarslan. Large-scale hydrogen energy storage in salt caverns. *International Journal of Hydrogen Energy*, 37(19):14265–14277, 2012.

[291] B Ozturk and D Yilmaz. Absorptive removal of volatile organic compounds from flue gas streams. *Process Safety and Environmental Protection*, 84(5):391–398, 2006.

[292] Surya Padinjarekutt, Huazheng Li, Shoujie Ren, Pranav Ramesh, Fanglei Zhou, Shiguang Li, Georges Belfort, and Miao Yu. Na+-gated nanochannel membrane for highly selective ammonia (nh3) separation in the haber-bosch process. *Chemical Engineering Journal*, 454:139998, 2023.

[293] Jose Palomar, Maria Gonzalez-Miquel, Jorge Bedia, Francisco Rodriguez, and Juan J Rodriguez. Task-specific ionic liquids for efficient ammonia absorption. *Separation and Purification Technology*, 82:43–52, 2011.

[294] Matthew J Palys. *Renewable ammonia for sustainable agriculture and energy: Process, system, and enterprise optimization*. PhD thesis, University of Minnesota, 2021.

[295] Matthew J Palys and Prodromos Daoutidis. Using hydrogen and ammonia for renewable energy storage: A geographically comprehensive techno-economic study. *Computers & Chemical Engineering*, 136:106785, 2020.

[296] Matthew J Palys and Prodromos Daoutidis. Power-to-x: A review and perspective. *Computers & Chemical Engineering*, 165:107948, 2022.

[297] Matthew J Palys, Anatoliy Kuznetsov, Joel Tallaksen, Michael Reese, and Prodromos Daoutidis. A novel system for ammonia-based sustainable energy and agriculture: Concept and design optimization. *Chemical Engineering and Processing-Process Intensification*, 140:11–21, 2019.

[298] Matthew J Palys, Ilias Mitrai, and Prodromos Daoutidis. Renewable hydrogen and ammonia for combined heat and power systems in remote locations: Optimal design and scheduling. *Optimal Control Applications & Methods*, 44(2):719–738, 2023.

[299] Dionissios D Papadias, Jui-Kun Peng, and Rajesh K Ahluwalia. Hydrogen carriers: Production, transmission, decomposition, and storage. *International Journal of Hydrogen Energy*, 46(47):24169–24189, 2021.

[300] Sameer Parvathikar, Ignacio Luz, Michael Carpenter, Timothy Bellamy, Kelly Amato, John Carpenter, Dennis Gilmore, and Marty Lail. Solvothermal synthesis of mof-derived supported ru nanocatalysts for low-temperature ammonia synthesis. *Catalysis Today*, 387:23–27, 2022.

[301] Venkat Pattabathula and Jim Richardson. Introduction to ammonia production. *Chemical Engineering Progress*, 112(9):69–75, 2016.

[302] Matthew A Pellow, Christopher JM Emmott, Charles J Barnhart, and Sally M Benson. Hydrogen or batteries for grid storage? A net energy analysis. *Energy & Environmental Science*, 8(7):1938–1952, 2015.

[303] Xuanbei Peng, Xiaochuan Chen, Yanliang Zhou, Fuxiang Sun, Tianhua Zhang, Lirong Zheng, Lilong Jiang, and Xiuyun Wang. Size-dependent activity of supported ru catalysts for ammonia synthesis at mild conditions. *Journal of Catalysis*, 408:98–108, 2022.

[304] Camille Petit, Mykola Seredych, and Teresa J Bandosz. Revisiting the chemistry of graphite oxides and its effect on ammonia adsorption. *Journal of Materials Chemistry*, 19(48):9176–9185, 2009.

[305] William A Phillip, Eddie Martono, Liang Chen, Marc A Hillmyer, and Edward L Cussler. Seeking an ammonia selective membrane based on nanostructured sulfonated block copolymers. *Journal of Membrane Science*, 337(1-2):39–46, 2009.

[306] Progressive Farmer powered by DTN. DTN Retail Fertilizer Trends. dtnpf.com/agriculture/web/ag/crops, 2024.

[307] Manoj Ravi and Joshua W Makepeace. Facilitating green ammonia manufacture under milder conditions: what do heterogeneous catalyst formulations have to offer? *Chemical Science*, 13(4):890–908, 2022.

[308] Michael Reese, Cory Marquart, Mahdi Malmali, Kevin Wagner, Eric Buchanan, Alon McCormick, and Edward L Cussler. Performance of a small-scale Haber process. *Industrial & Engineering Chemistry Research*, 55(13):3742–3750, 2016.

[309] Renewable Watch. AES Andes submits EIA for Inna green hydrogen project in Chile. renewablewatch.in/2025/01/02/aes-andes-submits-eia-for-inna-green-hydrogen-project-in-chile/, 2024.

[310] Simon Richard, Alvaro Ramirez Santos, Pierre Olivier, and Fausto Gallucci. Techno-economic analysis of ammonia cracking for large scale power generation. *International Journal of Hydrogen Energy*, 71:571–587, 2024.

[311] Adam J Rieth, Ashley M Wright, and Mircea Dincă. Kinetic stability of metal–organic frameworks for corrosive and coordinating gas capture. *Nature Reviews. Materials*, 4(11):708–725, 2019.

[312] Benjamin P Riley, Prodromos Daoutidis, and Qi Zhang. Multi-scenario design of ammonia-based energy storage systems for use as non-wires alternatives. *Journal of Energy Storage*, 73:108795, 2023.

[313] Kevin HR Rouwenhorst, Fatme Jardali, Annemie Bogaerts, and Leon Lefferts. From the birkeland-eyde process towards energy-efficient plasma-based no_x synthesis: a techno-economic analysis. *Energy & Environmental Science*, 14(5):2520–2534, 2021.

[314] Kevin HR Rouwenhorst, Anthony S Travis, and Leon Lefferts. 1921–2021: A century of renewable ammonia synthesis. *Sustainable Chemistry*, 3(2):149–171, 2022.

[315] Naghi Saadatjou, Ali Jafari, and Saeed Sahebdelfar. Ruthenium nanocatalysts for ammonia synthesis: a review. *Chemical Engineering Communications*, 202(4):420–448, 2015.

[316] Jhuma Sadhukhan, Kok Siew Ng, and Elias Martinez Hernandez. *Biorefineries and chemical processes: design, integration and sustainability analysis*. John Wiley & Sons, 2014.

[317] Victor N Sagel, Kevin HR Rouwenhorst, and Jimmy A Faria. Green ammonia enables sustainable energy production in small island developing states: A case study on the island of curaçao. *Renewable and Sustainable Energy Reviews*, 161:112381, 2022.

[318] Victor N Sagel, Kevin HR Rouwenhorst, and Jimmy A Faria. Renewable electricity generation in small island developing states: The effect of importing ammonia. *Energies*, 15(9):3374, 2022.

[319] Tetsuo Sakai, Hiroyasu Takenaka, and Eiichi Torikai. Gas diffusion in the dried and hydrated nafions. *Journal of the Electrochemical Society*, 133(1):88, 1986.

[320] Billur Sakintuna, Farida Lamari-Darkrim, and Michael Hirscher. Metal hydride materials for solid hydrogen storage: a review. *International Journal of Hydrogen Energy*, 32(9):1121–1140, 2007.

[321] Nicholas Salmon and René Bañares-Alcántara. Impact of grid connectivity on cost and location of green ammonia production: Australia as a case study. *Energy & Environmental Science*, 14(12):6655–6671, 2021.

[322] Nicholas Salmon and René Bañares-Alcántara. A global, spatially granular techno-economic analysis of offshore green ammonia production. *Journal of Cleaner Production*, 367:133045, 2022.

[323] Antonio Sánchez, Elena Castellano, Mariano Martín, and Pastora Vega. Evaluating ammonia as green fuel for power generation: A thermo-chemical perspective. *Applied Energy*, 293:116956, 2021.

[324] Antonio Sánchez, Mariano Martín, and Qi Zhang. Optimal design of sustainable power-to-fuels supply chains for seasonal energy storage. *Energy*, 234:121300, 2021.

[325] Antonio Sánchez, Qi Zhang, Mariano Martín, and Pastora Vega. Towards a new renewable power system using energy storage: An economic and social analysis. *Energy Conversion and Management*, 252:115056, 2022.

[326] Antonio Sánchez, Mariano Martín, and Pastora Vega. Biomass based sustainable ammonia production: digestion vs gasification. *ACS Sustainable Chemistry & Engineering*, 7(11):9995–10007, 2019.

[327] Robert Schlögl. Catalytic synthesis of ammonia—a "never-ending story"? *Angewandte Chemie International Edition*, 42(18):2004–2008, 2003.

[328] O Schmidt, A Gambhir, I Staffell, A Hawkes, J Nelson, and S Few. Future cost and performance of water electrolysis: An expert elicitation study. *International Journal of Hydrogen Energy*, 42(52):30470–30492, 2017.

[329] Jay Schmuecker. The carbon-free farm. *IEEE Spectrum*, 56(11):30–35, 2019.

[330] E Kilburn Scott. Nitrates and ammonia from atmospheric nitrogen. lectures iii. *Journal of the Royal Society of Arts*, 71(3704):900–917, 1923.

[331] Youngkyun Seo and Seongjong Han. Economic evaluation of an ammonia-fueled ammonia carrier depending on methods of ammonia fuel storage. *Energies*, 14(24):8326, 2021.

[332] Nestor A Sepulveda, Jesse D Jenkins, Aurora Edington, Dharik S Mallapragada, and Richard K Lester. The design space for long-duration energy storage in decarbonized power systems. *Nature Energy*, pages 1–11, 2021. doi: https://doi.org/10.1038/s41560-021-00796-8.

[333] Mykola Seredych and Teresa J Bandosz. Mechanism of ammonia retention on graphite oxides: Role of surface chemistry and structure. *The Journal of Physical Chemistry C*, 111(43):15596–15604, 2007.

[334] Mykola Seredych, Camille Petit, Albert V Tamashausky, and Teresa J Bandosz. Role of graphite precursor in the performance of graphite oxides as ammonia adsorbents. *Carbon*, 47(2):445–456, 2009.

[335] Vahid Shadravan, Ang Cao, Vanessa J Bukas, Mette K Grønborg, Christian D Damsgaard, Zhenbin Wang, Jakob Kibsgaard, Jens K Nørskov, and Ib Chorkendorff. Enhanced promotion of ru-based ammonia catalysts by in situ dosing of cs. *Energy & Environmental Science*, 15(8):3310–3320, 2022.

[336] Md Fahim Shahriar and Aaditya Khanal. The current techno-economic, environmental, policy status and perspectives of sustainable aviation fuel (saf). *Fuel*, 325:124905, 2022.

[337] Dawei Shang, Xiangping Zhang, Shaojuan Zeng, Kun Jiang, Hongshuai Gao, Haifeng Dong, Qingyuan Yang, and Suojiang Zhang. Protic ionic liquid [bim][ntf 2] with strong hydrogen bond donating ability for highly efficient ammonia absorption. *Green Chemistry*, 19(4):937–945, 2017.

[338] Jie Shi, Yuanqing Zhu, Yongming Feng, Jun Yang, and Chong Xia. A prompt decarbonization pathway for shipping: Green hydrogen, ammonia, and methanol production and utilization in marine engines. *Atmosphere*, 14(3):584, 2023.

[339] Sensen Shi, Yuanmeng Duan, Guoxiong Zhan, Fei Chang, Shaojuan Zeng, Xiaoyan Ji, Zhichang Liu, Xianghai Meng, and Xiangping Zhang. Simulation and energetic assessment of the ammonia synthesis loop with ionic liquid-based ammonia recovery from recycle gas. *Separation and Purification Technology*, 301:121951, 2022.

[340] Wei Shi and Edward J Maginn. Molecular simulation of ammonia absorption in the ionic liquid 1-ethyl-3-methylimidazolium bis (trifluoromethylsulfonyl) imide ([emim][tf2n]). *AIChE Journal*, 55(9):2414–2421, 2009.

[341] Naohiro Shimoda, Yutaka Kimura, Yusuke Kobayashi, Jun Kubota, and Shigeo Satokawa. Ammonia synthesis over yttrium-doped barium zirconate and cerate-based perovskite-type oxide supported ruthenium catalysts. *International Journal of Hydrogen Energy*, 42(50):29745–29755, 2017.

[342] David S Sholl and Ryan P Lively. Seven chemical separations to change the world. *Nature*, 532(7600):435–437, 2016.

[343] Irena Short, Ashok Sahgal, and Walter Hayduk. Solubility of ammonia and hydrogen sulfide in several polar solvents. *Journal of Chemical and Engineering Data*, 28(1):63–66, 1983.

[344] Vaclav Smil. Detonator of the population explosion. *Nature*, 400(6743):415, 1999.

[345] Vaclav Smil. Nitrogen and food production: proteins for human diets. *AMBIO: A Journal of the Human Environment*, 31(2):126–131, 2002.

[346] AR Smith and J Klosek. A review of air separation technologies and their integration with energy conversion processes. *Fuel Processing Technology*, 70(2):115–134, 2001.

[347] Collin Smith, Alfred K Hill, and Laura Torrente-Murciano. Current and future role of haber–bosch ammonia in a carbon-free energy landscape. *Energy & Environmental Science*, 13(2):331–344, 2020.

[348] Collin Smith and Laura Torrente-Murciano. Exceeding single-pass equilibrium with integrated absorption separation for ammonia synthesis using renewable energy—redefining the haber-bosch loop. *Advanced Energy Materials*, 11(13):2003845, 2021.

[349] Collin Smith and Laura Torrente-Murciano. The potential of green ammonia for agricultural and economic development in sierra leone. *One Earth*, 4(1):104–113, 2021.

[350] Benjamin ER Snyder, Ari B Turkiewicz, Hiroyasu Furukawa, Maria V Paley, Ever O Velasquez, Matthew N Dods, and Jeffrey R Long. A ligand insertion mechanism for cooperative nh3 capture in metal–organic frameworks. *Nature*, 613(7943):287–291, 2023.

[351] Rasmus Z Sørensen, Jens S Hummelshøj, Asbjørn Klerke, Jacob Birke Reves, Tejs Vegge, Jens K Nørskov, and Claus H Christensen. Indirect, reversible high-density hydrogen storage in compact metal ammine salts. *Journal of the American Chemical Society*, 130(27):8660–8668, 2008.

[352] Joana Sousa, Wendelin Waiblinger, and Kaspar Andreas Friedrich. Techno-economic study of an electrolysis-based green ammonia production plant. *Industrial & Engineering Chemistry Research*, 61(39):14515–14530, 2022.

[353] Claudia Staudt-Bickel and William J Koros. Improvement of co2/ch4 separation characteristics of polyimides by chemical crosslinking. *Journal of Membrane Science*, 155(1):145–154, 1999.

[354] G Kh Step and MV Petrovichev. Pressure swing adsorption for air separation and purification. *Chemical and Petroleum Engineering*, 38(3):154–158, 2002.

[355] K Stephan. History of absorption heat pumps and working pair developments in europe. *International Journal of Refrigeration*, 6(3):160–166, 1983.

[356] Joel Tallaksen. Life Cycle Assessment and Cropping Energy Audits for IREE Project RL-0016-13. Technical report, West Central Research and Outreach Center, University of Minnesota, Morris, MN, United States, 2017.

[357] TalusAg. Kenya Nut Company Installs First Commercial On-Site Green Ammonia System for Carbon-Free Fertilizer Production. talusag.com/blog/kenya-nut-company-installs-first-commercial-on-site-green-ammonia-system-for-carbon-free-fertilizer-production, 2023.

[358] Ya Tang, Yoji Kobayashi, Naoya Masuda, Yoshinori Uchida, Hiroki Okamoto, Toki Kageyama, Saburo Hosokawa, François Loyer, Kei Mitsuhara, Keisuke Yamanaka, et al. Metal-dependent support effects of oxyhydride-supported ru, fe, co catalysts for ammonia synthesis. *Advanced Energy Materials*, 8(36):1801772, 2018.

[359] Adem Tekin, Jens S Hummelshøj, Hjalte S Jacobsen, Dadi Sveinbjörnsson, Didier Blanchard, Jens K Nørskov, and Tejs Vegge. Ammonia dynamics in magnesium ammine from dft and neutron scattering. *Energy & Environmental Science*, 3(4):448–456, 2010.

[360] SF Timashev, AV Vorobiev, VI Kirichenko, Yu M Popkov, VI Volkov, RR Shifrina, A Ya Lyapunov, AG Bondarenko, and LP Bobrova. Specifics of highly selective ammonia transport through gas-separating membranes based on perfluorinated copolymer in the form of hollow fibers. *Journal of Membrane Science*, 59(2):117–131, 1991.

[361] Topsoe. Reduce carbon emissions with green ammonia. info.topsoe.com/green-ammonia, 2023.

[362] David Toyne and Jay Schmuecker. Our Iowa renewable hydrogen and ammonia generation system. In *14th Annual NH3 Fuel Conference: Enabling Optimized, Sustainable Energy and Agriculture*. NH_3 Fuel Association, 2017. nh3fuelassociation.org/wp-content/uploads/2017/11/NH3-Energy-2017-David-Toyne.pdf.

[363] David Toyne and Jay Schmuecker. Our Improved Farm Tractor Ammonia and Hydrogen Fueling System. ammoniaenergy.org/presentations/our-improved-farm-tractor-ammonia-and-hydrogen-fueling-system/, 2019.

[364] Anthony S Travis. Globalising synthetic nitrogen: the interwar inauguration of a new industry. *ambix*, 64(1):1–28, 2017.

[365] Anthony S Travis. *Nitrogen Capture: The Growth of an International Industry (1900–1940)*. Springer Chem, 2018.

[366] William W Tso, C Doga Demirhan, Seungyeon Lee, Haneol Song, Joseph B Powell, and Efstratios N Pistikopoulos. Energy carrier supply chain optimization: A Texas case study. In *Computer Aided Chemical Engineering*, volume 47, pages 1–6. Elsevier, 2019.

[367] Taku Tsujimura and Norihiko Iki (National Institute of Advanced Industrial Science and Technology). Gas turbine power generation with a methane-ammonia gas mixture and 100 % ammonia. aist.go.jp/aist_e/list/latest_research/2016/20160412/en20160412.html, 2016.

[368] Per Tunå, Christian Hulteberg, and Serina Ahlgren. Techno-economic assessment of nonfossil ammonia production. *Environmental Progress & Sustainable Energy*, 33(4):1290–1297, 2014.

[369] Kazuo Urabe, Ken-Ichi Aika, and Atsumu Ozaki. Activation of nitrogen by alkali metal-promoted transition metal: Vi. hydrogen effect on isotopic equilibration of nitrogen and rate-determining step of ammonia synthesis on potassium-promoted ruthenium catalysts. *Journal of Catalysis*, 42(2):197–204, 1976.

[370] U.S. Department of the Energy. National Transmission Needs Study. energy.gov/gdo/national-transmission-needs-study, 2023.

[371] U.S. Department of the Treasury Internal Revenue Service. Credit for Production of Clean Hydrogen and Energy Credit - Final Regulations. federalregister.gov/documents/2025/01/10/2024-31513/credit-for-production-of-clean-hydrogen-and-energy-credit, 2025.

[372] U.S. Energy Information Administration. Annual Energy Outlook 2021. Technical report, U.S. Department of Energy, Washington, DC (United States), 2021. eia.gov/outlooks/aeo.

[373] U.S. Energy Information Administration. Levelized Costs of New Generation Resources in the Annual Energy Outlook 2022. eia.gov/outlooks/aeo/pdf/electricity_generation.pdf, 2022.

[374] U.S. Energy Information Administration. Henry Hub Natural Gas Spot Price. eia.gov/dnav/ng/hist/rngwhhdA.htm, 2025.

[375] U.S. Energy Information Administration - Independent Statistics and Analysis. Biomass and waste electricity net generation. eia.gov/international/data/world/electricity/electricity-generation, 2024.

[376] U.S. Energy Information Administration - Independent Statistics and Analysis. World biofuels production. eia.gov/international/data/world/biofuels/more-biofuels-data, 2024.

[377] Agustin Valera-Medina and Rene Banares-Alcantara. *Techno-economic challenges of green ammonia as an energy vector*. Academic Press, 2020.

[378] Agustin Valera-Medina, H Xiao, M Owen-Jones, WIF David, and PJ Bowen. Ammonia for power. *Progress in Energy and Combustion Science*, 69:63–102, 2018.

[379] Tejs Vegge, Rasmus Zink Sørensen, Asbjørn Klerke, Jens Strabo Hummelshøj, T Johannessen, Jens Kehlet Nørskov, and CH Christensen. Indirect hydrogen storage in metal ammines. In *Solid-State Hydrogen Storage*, pages 533–564. Elsevier, 2008.

[380] Kevin Verleysen, Diederik Coppitters, Alessandro Parente, and Francesco Contino. Where to build the ideal solar-powered ammonia plant? design optimization of a belgian and moroccan power-to-ammonia plant for covering the belgian demand under uncertainties. *Applications in Energy and Combustion Science*, 14:100141, 2023.

[381] Kumar Vikrant, Vanish Kumar, Ki-Hyun Kim, and Deepak Kukkar. Metal–organic frameworks (mofs): potential and challenges for capture and abatement of ammonia. *Journal of Materials Chemistry A*, 5(44):22877–22896, 2017.

[382] Tymen Visser, N Masetto, and Matthias Wessling. Materials dependence of mixed gas plasticization behavior in asymmetric membranes. *Journal of Membrane Science*, 306(1-2):16–28, 2007.

[383] Tymen Visser and Matthias Wessling. When do sorption-induced relaxations in glassy polymers set in? *Macromolecules*, 40(14):4992–5000, 2007.

[384] Aleksandra Vojvodic, Andrew James Medford, Felix Studt, Frank Abild-Pedersen, Tuhin Suvra Khan, T Bligaard, and JK Nørskov. Exploring the limits: A low-pressure, low-temperature haber–bosch process. *Chemical Physics Letters*, 598:108–112, 2014.

[385] Kevin Wagner, Mahdi Malmali, Collin Smith, Alon McCormick, EL Cussler, Ming Zhu, and Nicholas CA Seaton. Column absorption for reproducible cyclic separation in small scale ammonia synthesis. *AIChE Journal*, 63(7):3058–3068, 2017.

[386] Stanley M Walas. *Chemical Process Equipment: Selection and Design*. Elsevier, 2013.

[387] Changlong Wang, Stuart DC Walsh, Thomas Longden, Graham Palmer, Israel Lutalo, and Roger Dargaville. Optimising renewable generation configurations of off-grid green ammonia production systems considering haber-bosch flexibility. *Energy Conversion and Management*, 280:116790, 2023.

[388] Ganzhou Wang, Alexander Mitsos, and Wolfgang Marquardt. Conceptual design of ammonia-based energy storage system: System design and time-invariant performance. *AIChE Journal*, 63(5):1620–1637, 2017.

[389] Hanchu Wang, Prodromos Daoutidis, and Qi Zhang. Harnessing the wind power of the ocean with green offshore ammonia. *ACS Sustainable Chemistry & Engineering*, 9(43):14605–14617, 2021.

[390] Hanchu Wang, Prodromos Daoutidis, and Qi Zhang. Ammonia-based green corridors for sustainable maritime transportation. *Digital Chemical Engineering*, 6:100082, 2023.

[391] Hanchu Wang, Matthew Palys, Prodromos Daoutidis, and Qi Zhang. Optimal design of sustainable ammonia-based food–energy–water systems with nitrogen management. *ACS Sustainable Chemistry & Engineering*, 9(7):2816–2834, 2021.

[392] Junli Wang, Shaojuan Zeng, Feng Huo, Dawei Shang, Hongyan He, Lu Bai, Xiangping Zhang, and Jianwei Li. Metal chloride anion-based ionic liquids for efficient separation of nh3. *Journal of Cleaner Production*, 206:661–669, 2019.

[393] Peikun Wang, Fei Chang, Wenbo Gao, Jianping Guo, Guotao Wu, Teng He, and Ping Chen. Breaking scaling relations to achieve low-temperature ammonia synthesis through lih-mediated nitrogen transfer and hydrogenation. *Nature Chemistry*, 9(1):64–70, 2017.

[394] Qinghua Wang, Yun Wang, Xiaoxue Sun, Li Wei, Ligang Wei, Shangru Zhai, Zuoyi Xiao, Qingda An, and Jingai Hao. Constructing ternary deep eutectic solvents with multiple sites for ammonia storage. *International Journal of Hydrogen Energy*, 47(80):34102–34111, 2022.

[395] Tongzhou Wang, Xuejie Cao, and Lifang Jiao. Pem water electrolysis for hydrogen production: fundamentals, advances, and prospects. *Carbon Neutrality*, 1(1):21, 2022.

[396] Xiao Wang, Lin Guo, Jing Lv, Maoshuai Li, Shouying Huang, Yue Wang, and Xinbin Ma. Process design, modeling and life cycle analysis of energy consumption and ghg emission for jet fuel production from bioethanol in china. *Journal of Cleaner Production*, 389:136027, 2023.

[397] Xiuyun Wang, Xuanbei Peng, Wei Chen, Guangyong Liu, Anmin Zheng, Lirong Zheng, Jun Ni, Chak-tong Au, and Lilong Jiang. Insight into dynamic and steady-state active sites for nitrogen activation to ammonia by cobalt-based catalyst. *Nature Communications*, 11(1):653, 2020.

[398] Ziqing Wang, Jianxin Lin, Rong Wang, and Kemei Wei. Ammonia synthesis over ruthenium catalyst supported on perovskite type batio3. *Catalysis Communications*, 32:11–14, 2013.

[399] Ziqing Wang, Benyao Liu, and Jianxin Lin. Highly effective perovskite-type bazro3 supported ru catalyst for ammonia synthesis. *Applied Catalysis. A, General*, 458:130–136, 2013.

[400] Qiushi Wei, Jolie M Lucero, James M Crawford, J Douglas Way, Colin A Wolden, and Moises A Carreon. Ammonia separation from n2 and h2 over lta zeolitic imidazolate framework membranes. *Journal of Membrane Science*, 623:119078, 2021.

[401] John Weiss, Tom Bruulsema, Mike Hunter, Karl Czymmek, Joe Lawrence, and Quirine Ketterings. Nitrogen Fertilizers for Field Crops. Technical report, Cornell University Field Crops Extension, Ithaca, NY (USA), 2009. nmsp.cals.cornell.edu/publications/factsheets/factsheet44.pdf.

[402] Tess Weiss, Chathurika Gamage, Thomas Koch Blank, Genevieve Lillis, and Jardine Wall. Hydrogen Reality Check: All “Clean Hydrogen” Is Not Equally Clean. rmi.org/all-clean-hydrogen-is-not-equally-clean/, 2022.

[403] Jiazhen Wu, Yutong Gong, Takeshi Inoshita, Daniel C Fredrickson, Junjie Wang, Yangfan Lu, Masaaki Kitano, and Hideo Hosono. Tiered electron anions in multiple voids of lascsi and their applications to ammonia synthesis. *Advanced Materials*, 29(36):1700924, 2017.

[404] Jiazhen Wu, Jiang Li, Yutong Gong, Masaaki Kitano, Takeshi Inoshita, and Hideo Hosono. Intermetallic electride catalyst as a platform for ammonia synthesis. *Angewandte Chemie International Edition*, 58(3):825–829, 2019.

[405] Shengwei Wu, Bin Miao, and Siew Hwa Chan. Feasibility assessment of a container ship applying ammonia cracker-integrated solid oxide fuel cell technology. *International Journal of Hydrogen Energy*, 47(63):27166–27176, 2022.

[406] Yan Wu, Tingting Zhao, Shuo Tang, Yufei Wang, and Mingyan Ma. Research on design and multi-frequency scheduling optimization method for flexible green ammonia system. *Energy Conversion and Management*, 300:117976, 2024.

[407] Yuyuan Wu, Chunyan Li, Biyun Fang, Xiuyun Wang, Jun Ni, Bingyu Lin, Jianxin Lin, and Lilong Jiang. Enhanced ammonia synthesis performance of ceria-supported ru catalysts via introduction of titanium. *Chemical Communications*, 56(7):1141–1144, 2020.

[408] Bingbing Yang, Lu Bai, Tingting Li, Liyuan Deng, Lei Liu, Shaojuan Zeng, Jiuli Han, and Xiangping Zhang. Super selective ammonia separation through multiple-site interaction with ionic liquid-based hybrid membranes. *Journal of Membrane Science*, 628:119264, 2021.

[409] Mengyao Yang and Jasmine Siu Lee Lam. Operational and economic evaluation of ammonia bunkering–bunkering supply chain perspective. *Transportation Research. Part D, Transport and Environment*, 117:103666, 2023.

[410] Arda Yapicioglu and Ibrahim Dincer. A review on clean ammonia as a potential fuel for power generators. *Renewable and Sustainable Energy Reviews*, 103:96–108, 2019.

[411] Tian-Nan Ye, Sang-Won Park, Yangfan Lu, Jiang Li, Masato Sasase, Masaaki Kitano, Tomofumi Tada, and Hideo Hosono. Vacancy-enabled n2 activation for ammonia synthesis on an ni-loaded catalyst. *Nature*, 583(7816):391–395, 2020.

[412] A Yokozeki and Mark B Shiflett. Ammonia solubilities in room-temperature ionic liquids. *Industrial & Engineering Chemistry Research*, 46(5):1605–1610, 2007.

[413] Zhipeng Yu, Jin Lin, Feng Liu, Jiarong Li, Yuxuan Zhao, and Yonghua Song. Optimal sizing of isolated renewable power systems with ammonia synthesis: Model and solution approach. *IEEE Transactions on Power Systems*, 39:6372–6385, 2024.

[414] Lei Yuan, Xiangping Zhang, Baozeng Ren, Yingliang Yang, Yinge Bai, Lu Bai, Hongshuai Gao, and Shaojuan Zeng. Dual-functionalized protic ionic liquids for efficient absorption of nh3 through synergistically physicochemical interaction. *Journal of Chemical Technology and Biotechnology*, 95(6):1815–1824, 2020.

[415] C Zamfirescu and I Dincer. Using ammonia as a sustainable fuel. *Journal of Power Sources*, 185(1):459–465, 2008.

[416] Kai Zeng and Dongke Zhang. Recent progress in alkaline water electrolysis for hydrogen production and applications. *Progress in Energy and Combustion Science*, 36(3):307–326, 2010.

[417] Hanfei Zhang, Ligang Wang, François Maréchal, Umberto Desideri, et al. Techno-economic comparison of green ammonia production processes. *Applied Energy*, 259:114135, 2020.

[418] Jian-Bo Zhang and Kuan Huang. Solubilities of ammonia in polyethylene glycols at 298.2–353.2 k and 0–200 kpa. *Journal of Chemical and Engineering Data*, 65(1):97–105, 2019.

[419] Jiayin Zhang, Lu Zheng, Yongde Ma, Zhenping Cai, Yanning Cao, Kuan Huang, and Lilong Jiang. A mini-review on nh_3 separation technologies: recent advances and future directions. *Energy & Fuels*, 36(24):14516–14533, 2022.

[420] Lu Zhang, Haifeng Dong, Shaojuan Zeng, Zongyuan Hu, Shahid Hussain, and Xiangping Zhang. An overview of ammonia separation by ionic liquids. *Industrial & Engineering Chemistry Research*, 60(19):6908–6924, 2021.

[421] Yuqi Zhang, Ning Zhao, Mingqiang Li, Zhan Xu, Dawei Wu, Stuart Hillmansen, Athanasios Tsolakis, Kevin Blacktop, and Clive Roberts. A techno-economic analysis of ammonia-fuelled powertrain systems for rail freight. *Transportation Research. Part D, Transport and Environment*, 119:103739, 2023.

[422] Zhenyu Zhang, J Douglas Way, and Colin A Wolden. Design and operational considerations of catalytic membrane reactors for ammonia synthesis. *AIChE Journal*, 67(8):e17259, 2021.

[423] Fei Zhao, Yalou Li, Xiaoxin Zhou, Dandan Wang, Yawei Wei, and Fang Li. Co-optimization of decarbonized operation of coal-fired power plants and seasonal storage based on green ammonia co-firing. *Applied Energy*, 341:121140, 2023.

[424] Hong-Yong Zhao, Yi-Ming Cao, Xiao-Li Ding, Mei-Qing Zhou, Jian-Hui Liu, and Quan Yuan. Poly (ethylene oxide) induced cross-linking modification of matrimid membranes for selective separation of co2. *Journal of Membrane Science*, 320(1-2):179–184, 2008.

[425] Jiemin Zhao, Jiande Zhou, Mingwei Yuan, and Zhixiong You. Controllable synthesis of ru nanocrystallites on graphene substrate as a catalyst for ammonia synthesis. *Catalysis Letters*, 147:1363–1370, 2017.

[426] Jianwei Zheng, Fenglin Liao, Simson Wu, Glenn Jones, Tian-Yi Chen, Joshua Fellowes, Tim Sudmeier, Ian J McPherson, Ian Wilkinson, and Shik Chi Edman Tsang. Efficient non-dissociative activation of dinitrogen to ammonia over lithium-promoted ruthenium nanoparticles at low pressure. *Angewandte Chemie International Edition*, 58(48):17335–17341, 2019.

[427] Fu-Yu Zhong, Hai-Long Peng, Duan-Jian Tao, Ping-Keng Wu, Jie-Ping Fan, and Kuan Huang. Phenol-based ternary deep eutectic solvents for highly efficient and reversible absorption of nh3. *ACS Sustainable Chemistry & Engineering*, 7(3):3258–3266, 2019.

[428] Fu-Yu Zhong, Linsen Zhou, Jian Shen, Yong Liu, Jie-Ping Fan, and Kuan Huang. Rational design of azole-based deep eutectic solvents for highly efficient and reversible capture of ammonia. *ACS Sustainable Chemistry & Engineering*, 7(16):14170–14179, 2019.

[429] Zhi-hua Zhong and Ken-ichi Aika. Effect of ruthenium precursor on hydrogen-treated active carbon supported ruthenium catalysts for ammonia synthesis. *Inorganica Chimica Acta*, 280(1-2):183–188, 1998.

[430] Jiahui Zhou, Ziyue Zhang, Runzhi Zhang, Wentao Zhang, Gang Xu, and Haiming Wang. Optimal capacity and multi-stable flexible operation strategy of green ammonia systems: Adapting to fluctuations in renewable energy. *Energy Conversion and Management*, 314:118720, 2024.

[431] Yanliang Zhou, Congying Wang, Xuanbei Peng, Tianhua Zhang, Xiuyun Wang, Yafei Jiang, Haifeng Qi, Lirong Zheng, Jianxin Lin, and Lilong Jiang. Boosting efficient ammonia synthesis over atomically dispersed co-based catalyst via the modulation of geometric and electronic structures. *CCS Chemistry*, 4(5):1758–1769, 2022.

[432] Yanliang Zhou, Cong-Qiao Xu, Zhenni Tan, Hongfang Cai, Xiuyun Wang, Jialiang Li, Lirong Zheng, Chak-tong Au, Jun Li, and Lilong Jiang. Integrating dissociative and associative routes for efficient ammonia synthesis over a ticn-promoted ru-based catalyst. *ACS Catalysis*, 12(4):2651–2660, 2022.

[433] Yaping Zhou, Yongcheng Ma, Guojun Lan, Haodong Tang, Wenfeng Han, Huazhang Liu, and Ying Li. A highly stable and active mesoporous ruthenium catalyst for ammonia synthesis prepared by a rucl3/sio2-templated approach. *Chinese Journal of Catalysis*, 40(1):114–123, 2019.

[434] Micah S Ziegler, Joshua M Mueller, Gonçalo D Pereira, Juhyun Song, Marco Ferrara, Yet-Ming Chiang, and Jessika E Trancik. Storage requirements and costs of shaping renewable energy toward grid decarbonization. *Joule*, 3(9):2134–2153, 2019.

Index

https://doi.org/10.1515/9783110762525-010

Printed and bound by CPI Group (UK) Ltd, Croydon, CR0 4YY
16/09/2026
14963129-0002